Electrical Safety
Engineering

Electrical Safety Engineering

W. FORDHAM COOPER

B.Sc., C.Eng., F.I.E.E., F.I.Mech.E

Former HM Electrical Inspector of Factories

NEWNES-BUTTERWORTHS
LONDON — BOSTON
Sydney - Wellington - Durban - Toronto

The Butterworth Group

United Kingdom   **Butterworth & Co (Publishers) Ltd.**
London: 88 Kingsway, WC2B 6AB

Australia   **Butterworths Pty Ltd**
Sydney: 586 Pacific Highway, Chatswood, NSW 2067
Also at Melbourne, Brisbane, Adelaide and Perth

Canada   **Butterworth & Co. (Canada) Ltd**
Toronto: 2265 Midland Avenue, Scarborough,
Ontario M1P 4S1

New Zealand   **Butterworths of New Zealand Ltd**
Wellington: 77–85 Customhouse Quay, 1

South Africa   **Butterworth & Co (South Africa) (Pty) Ltd**
Durban: 152–154 Gale Street

USA   **Butterworth (Publishers) Inc**
Boston: 19 Cummings Park, Woburn, Mass. 01801

First published 1978

© Butterworth & Co (Publishers) Ltd 1978

British Library Cataloguing in Publication Data

Cooper, W Fordham
  Electrical safety engineering.
  1. Electric engineering – Safety measures
  I. Title
  621.3   TK152   77–30158

  ISBN 0–408–00289–1

Typeset and produced by **Scribe Design** · Medway · Kent
Printed in England by Cox & Wyman Ltd.,
London, Fakenham and Reading

Preface

A few years ago I was asked by the Institution of Electrical Engineers to prepare a critical review of the safety aspects of electrical engineering. This was the starting point of this book, for although the review ran to about 35 000 words, it was possible to deal only very briefly with many aspects of this wide-ranging subject and some quite important matters were completely omitted. In this book I have attempted to make good some of these omissions in so far as they affect general industry. Electricity supply systems are not covered except incidentally; this would require a second volume.

Electrical engineering is an art as well as a branch of applied science and is continually developing. It follows that on many points there can be no final or definitive answers. In some cases opinions differ and it may be better to do the best one can at the time rather than to wait for perfection. Sometimes no satisfactory solution has yet been found and in these cases I have not hesitated to say so, at the same time giving the best answer.

A great deal of 'safety' must be built in at the design stage and cannot be dealt with effectively after manufacture and/or installation; this can apply both to plant costing a few pounds or many thousands of pounds. Safe design is therefore of as much concern to engineers and draughtsmen as safe practice is to works management and safety officers. It is for such people that this book has been written.

Some years ago I was asked to give some public lectures on this matter at a large technical college. Instead of the expected audience (of 25 or 30) 120 turned up. They included the senior engineers of local steel works, the gas and electricity supply industries, major manufacturers and so on — but *no students*. When questioned they said that the subject was elementary and they had better things to do. If, therefore, I can also persuade the departmental heads of universities and polytechnics that the subject is academically respectable, and as essential to the

future engineer in industry as a knowledge of the Gunn Effect, I shall have accomplished an important part of my purpose. I would also suggest that by the very diversity of the subject matter it could be made to serve as a bridge between mathematical analysis and practical applications, provide subjects for projects and introducing students to the practical limitations within which industry necessarily works.

To facilitate the use of this book as a work of reference, each chapter is independent although this entails some minor overlapping and repetition. To assist the reader considerable use has been made of cross-references throughout the text.

The Law requires that apparatus and procedures must be safe so far as is reasonably practicable (which does not mean reasonably convenient). However, what is practicable today may have been almost impossible 25 or 30 years ago, and what is safe in the hands of a highly skilled engineer may not be when used by unskilled workers. I have therefore attempted to deal with the underlying principles relating to the safe design and use of apparatus rather than details of hardware which may become obsolete. But some points of detail are included as examples when these are necessary to illustrate the principle.

When this book was first planned I intended to include a chapter on the legal aspects of accident prevention. However, the Health and Safety at Work Act has now come into force. Under this Act, a Health and Safety Commission and a Health and Safety Executive have been established, bringing together the various inspectorates concerned with Factories, Mines, Explosives, etc under one administration. The Act also considerably alters the powers of Inspectors and the means of enforcement and because it is an enabling Act, how matters will develop is, at present, uncertain. It is, however, clear that the range of safety legislation will be greatly extended since the scope of the Act relates to virtually all forms of employment and even, in some cases, to persons who are not employed. For this reason a comprehensive survey of the Act and other Legislation has been omitted, although the legal requirements are referred to where necessary in the text. So far as electrical installations in places covered by The Factories Act are concerned, the position is as yet unchanged on technical matters and the Electricity Regulations are for the present unaltered. Also, conditions and requirements vary from place to place and readers should always check the requirements of the latest edition of the various Regulations, Specifications and official recommendations both national and local. If a second edition of this book is required it is possible that a full account of the working of the Health and Safety Act may then be included.

SI units are used throughout the book, but in some cases where original experimental work has been quoted, the data has been kept in

the original units. Also, the practical ohm-cm unit of resistivity is used in a number of formulae, i.e. in relation to earthing.

Finally, this book has been written in the first person for two reasons. In the first place it is easier to avoid inflated circumlocutions and secondly it is based very largely on my own personal experience and investigations, although of course these were largely influenced by the work of colleagues; and it is, to that extent, at least, fallible and I do not wish to hide behind a mask of bogus authority.

W. Fordham Cooper

Contents

Acknowledgements

I should like to acknowledge the help of the following friends and colleagues who have assisted in the preparation of this book by reading and commenting on various parts of the text.

Professor Brian Harvey, late Chairman of the Health and Safety Executive for help with the chapters on General Principles and Calamity Hazards.

F. Arnaud, HM Superintending Technical Electrical Inspector of Factories and Mr. A. Hall, HM Superintending Electrical Inspector for their suggestions and comments on Chapters 9, 11 and 12.

K. Goodman, formerly HM Electrical Inspector for reading Chapter 9.

H. Vosper, Lecturer in Electrical Engineering, Brunel University for reading and commenting on Chapters, 2, 6, 7 and 8.

H.W. Turner of the ERA for information on fuses etc included in Chapter 7.

Dr. Black, formerly Chief Inspector of Explosives and Mr. Whitbread, Chief Inspector of Explosives.

Mr K.N. Palmer, Head of Fire Protection Division, Building Research Establishment and Dr C. Noble.

Messrs. A. Cooper and W.K. Dows for help with Chapters 3 and 4.

This book has been largely based on the following papers which I have written in the past and I wish to express my thanks to the editors and publishers for permission to quote from them. I should also like to thank the Controller of HM Stationery Office and the British Standards Institution for permission to quote from the various reports and official reports and British Standards.

'Magnetic fields surrounding a system of conductors (refers to short circuit forces) ERA, *World Power* (1925)

'The factor of safety of overhead lines', *The Electrician* (1929)

'The force of the wind on engineering structures', *Jour IEE* (1933)

'Safety and efficiency in electrical installation work', *Jour JIE* (1933)

'The electrical equipment of coke ovens and by-product works' and 'The planning of electrical maintenance and extensions', *Coke Oven Managers Assoc Year Book & Gas World*

'Control and protection of electrical furnaces', ASEE, *Electrical Supervisor* (1942)

'Load density on traction, power and domestic supplies (statistical)', Chairman's Address, *Jour IEE* (1938)

'The interpretation of statistical data', Chairman's Address, *Jour IEE* (1942)

'Insulating oil in relation to circuit breaker failures', *Jour IEE* (1943)

'Industrial fire risks' (with F.H. Mann), *Jour IEE* (1944)

'Electrical control of dangerous machinery and processes', *Jour IEE* (1947)

Part 2, 'Electrically interlocked guards on machines', *Jour IEE* (1951)

Part 3, 'Remote and supervisory control', *Jour IEE* (1953)

'Frictional electricity hazards in handling inflammable liquids', Abstract of lecture at symposium, *Trans. Inst of the Rubber Industry*, **23**

'The practical estimation of electrostatic hazards' and 'The electrification of fluids in motion', Inst of Physics, British Journal of Applied Physics Supplement No. 2 (1953)

'Mechanical and thermal failures in electrical switches and circuit breakers', Lecture to IEE in Sheffield and elsewhere

'Double insulation', *Electrical Times* (1964)

'High voltage switches and isolators', *Electrical Times* (1964)

'MV switches and isolators', *Electrical Times* (1964)

'Uses and abuses of fuses', *Electrical Times* (1965)

'Mechanical causes of electrical failures', *Electrical Times* (1966)

'Electrical equipment for areas of low explosive risk', ASEE, *Electrical Supervisor* (1967)

'Automatic fire detection systems', *Industrial Systems and Equipment* (1967)

'Electrical safety in industry', *IEE Review* (Aug 1970)

'Electrical safety engineering', Lecture to IEE Graduates, *Electronics and Power* (May 1973)

Abbreviations

IEE	Institution of Electrical Engineers
IMechE	Institution of Mechanical Engineers
BSI	British Standards Institution
ERA	Electrical Research Association
BASEEFA	British Approvals Service for Electrical Equipment in Flammable Atmospheres (Department of Trade & Industry)
BEAB	British Electrical Approvals Board (Testing and Certifying Authority for safety of domestic and similar electrical appliances)
ISO	International Standards Organisation
IEC	International Electrotechnical Commission (Standards Authority)
CENELEC	European Committee for Electrotechnical Standards
CEE	International Commission on Rules for the Approval of Electrical Equipment (Commission Internationale de réglementation en vue de l'approbation de l'équipement électrique) (Domestic and similar equipment only)
VDE	Verband Deutcher Elecktroteckniker (German Electrical Standards Authority)
PTB	Physikalish Technischen Bundesanstalt (German Certifying Authority)
BS	British Standard
CP	Code of Practice

SI   Statutory Instrument (general term for Statutory Rules, Regulations, etc)

HMSO   Her Majesty's Stationery Office

For convenience, the Electricity (Factories Act) Special Regulations 1908 and 1944 are referred to simply as the Electricity Regulations or Factory Electricity Regulations.

Chapter 1

Introduction

Some time ago I was asked to address the London Graduates and Students of the Institution of Electrical Engineers on the subject of Electrical Safety Engineering. That lecture formed the basis of this introductory chapter and explains why it is addressed primarily to young engineers. I hope, however, that all readers will find in it something of interest.

Electrical accidents, unlike most other industrial accidents, quite often happen to professional and supervisory staff. In some situations, they may be at greater risk than the manual staff. In a typical year 47% of electrical accidents involved electrically skilled persons and out of a total of 805 electrical accidents, 57 were to supervisory and testing staff (Table 1.2).

An engineer in charge of hazardous operations such as testing high-voltage apparatus in the field or supervising alterations in an important switching enclosure which cannot be made completely 'dead' has both legal and moral duties which he cannot, and should not wish to escape. He must conduct his work in accordance with Factory Regulations, and ensure that those under his control do the same. He must not do anything liable to endanger himself or anyone else. These are statutory obligations, quite distinct from liability for negligence in civil law. All engineers should therefore study the advisory *Memorandum on the Electricity Regulations*[1].

From time to time, we all make mistakes, but when life and limb are at risk it is inexcusable for anyone to take chances. It would be very sad to go through life knowing that someone had been killed or seriously injured owing to our own gross negligence.

In this book I have not attempted to cover electrical hazards which are common knowledge and present no problems; these are for the most part taken for granted. I have, however, concentrated on particular aspects which, from past experience, can cause trouble. The length and detail of the discussion is not necessarily an indication of relative importance. Some very important matters are straightforward and the

cure is obvious; others which are obscure and lead to controversy must be dealt with at greater length although they are relatively uncommon. Nearly all electrical accidents, even at low voltages (e.g. 230 V a.c.) are potentially fatal.

1.1 Control of staff and permits to work

No one should be allowed to do anything which is likely to be dangerous unless they have the necessary skill and experience and the technical knowledge to do the work safely. Managers and supervisors must therefore satisfy themselves that no one is asked to do any work for which they are not qualified.

Particular care must be taken with trainees and apprentices. Apprentices must learn to do potentially dangerous jobs during their apprenticeship. They must be carefully supervised while they are learning, and

Table 1.1 ELECTRICAL ACCIDENTS ANALYSED BY APPARATUS

	Fatal	*Total*
Portable tools (Class 1)	–	19
Heaters and irons	–	7
Lamps	2	13
Testing sets, including lamps, instruments and test leads	2	22
Plugs, sockets, couplings and adaptors	–	59
Cables and flex for portables (other than test sets)	2	33
Electric hand welding (excluding welding eye flash)	–	15
All other portable apparatus (including pendant controls)	–	24
Rotating electrical machines	–	13
Transformers and reactors	1	7
OCB's above 650 V	–	6
Oil immersed isolating switches above 650 V	–	9
Other switch, fuse and control gear above 650 V	–	9
Circuit breakers, not exceeding 650 V	–	17
Contactor and other control apparatus below 650 V	–	93
Switches and links not exceeding 650 V	1	76
Fuse gear not exceeding 650 V	2	58
Crane and other trolley wires, etc	1	16
Fixed lamps	1	14
Cables and accessories (excluding cables, flexibles etc for portable apparatus and buried cables)	–	–
Buried cables	–	75
Fixed test apparatus and their cables	–	8
Contact by cranes and similar machines	6	16
Direct contact by persons, materials, tools	4	16
Batteries	–	31
HF heating apparatus	1	9
Radio, TV, electronic instruments and power packs	–	11
Apparatus not classified	2	37
Total	26	805

Table 1.2 ELECTRICAL ACCIDENTS ANALYSED BY OCCUPATION

	Fatal	*Total*
SKILLED		
Supervisory staff	2	37
Switchboard substation attendants	–	2
Testing staff	–	18
Electrical tradesmen and their mates	7	278
Electrical engineering apprentices (under 18)	–	7
Electrical engineering apprentices (over 18)	1	17
UNSKILLED		
All men not included in the above	16	396
All women not included in the above	–	50
Total	26 (3.20%)	805 (100%)

they must not be too young because boys will be boys and they inevitably tend to lark about. For these reasons special provisions have been made to regulate their training and employment.

When there may be danger, and some electrical work (particularly in testing and maintenance) cannot be made absolutely safe, instructions must be very precise and unambiguous. Much work is done under permits to work, which must be issued and cancelled in an orderly and clearly defined manner. A full record must be kept so that it is possible, at any time, to find out what is going on, who is involved or at risk, and what precautions have been taken.

The permit must state clearly and fully to whom it is issued (this person should be present at all times and is responsible for what happens), name persons who may be present in the danger area, and state what special precautions have been taken to prevent danger. The safe and unsafe areas must be stated, and clearly indicated on the site. The work to be done must be clearly stated, and no other work must be carried out, because it may entail risks not contemplated by the person issuing the permit, and therefore may not have taken the necessary extra precautions.

At the end of the work there must be a clearly defined procedure for handing over. A check must be made that all persons have been withdrawn and the result recorded. A statement must be recorded, preferably on the back of the cancelled permit and in the logbook (where one is kept), of what work has been done – and what is left undone – and what steps have been taken to render the site safe for normal operations before the permit is cancelled. Until the permit has been cancelled the person to whom it was issued remains responsible for everything that happens.

If the work lasts for more than one shift there must be an appropriate method of handing over and ensuring that the new shift supervisor is familiar with the state of the work and the terms of the permit. It is often preferable to cancel the first permit and issue a new one. Sometimes the person with the authority to issue permits takes charge of the work; in that case he should issue a permit for himself.

All this detailed procedure may sound very fussy, but experience has shown that it is essential. The routine not only ensures that there is a record which should show the cause of any mistake, but the mere writing down of all the details is a great help in preventing anything being overlooked. As the persons concerned must sign all records and statements, the routine helps to insure that instructions have been read and understood.

Some testing and research work presents its own hazards. As the conditions may vary greatly, it is impossible to lay down general rules in detail, and safety depends largely on the skill of the staff. For certain work, because apparatus must be handled live, earth-free areas and unearthed tools such as soldering irons are provided. The soldering irons should, however, be supplied at a very low voltage.

Routine high-voltage testing is normally carried out in enclosures with interlocked doors and provision for supervision from the outside. There is an official pamphlet available dealing at some length with safe procedures[2]. Where unskilled or semiskilled persons do routine testing on a production line, arrangements must be made, by guards and interlocks for example, to ensure that they have access to live circuits only at very low voltages, say about 12 V.

Whoever is immediately responsible for the safety of a dangerous operation, whether that person is a senior engineer or a semi skilled 'watchman', must not leave the job or lend a hand in case their attention is distracted. These matters are discussed in the following chapter.

1.2 Non-electrical causes

Some 'so-called' electrical accidents are the result of mechanical and other causes. Examples of these causes are mechanical 'stress-raisers', thermal shock on insulators, resonant vibrations of conductors leading to fractures, low temperature brittleness or corrosion fatigue. To deal with such troubles it is necessary to have more than a narrow interest in electrical matters. Some notes on these aspects of the matter are included in Chapter 5.

The official report on the enquiry into the disastrous explosion and fire at Flixborough stated that engineers should have academic and practical training in all branches of engineering, other than their speciality, which may affect their work (see Appendix).

1.3 Detailed design

British Standards are necessarily very precise about small details of apparatus to ensure that it is not only safe when leaving the manufacturers but also remains safe in use and after repairs. Some of the conditions laid down may appear trivial, but are, in fact, essential. It is not easy to decide what to specify as to attain safety without limiting choice of design.

Some important basic requirements of British Standards are:

The insulation of conductors shall be unable to come into contact with moving parts.

Earthing terminals shall be adequately locked against loosening. These terminals shall not serve for any other purpose, e.g. for securing parts of the case.

Electrical connections shall be so designed that the contact pressure is not transmitted through insulating material other than ceramic or other materials not subject to shrinkage or deterioration.

Knobs, handles, operating levers and the like, which when removed or damaged render live joints accessible, shall be of adequate mechanical strength, and shall be so attached to the shaft that they cannot become detached inadvertently, even after extensive use. They shall be so arranged that contact with live shafts cannot be made by thin metal objects allowed the fall between the knobs and the case.

Soldered connections shall be so designed that they keep the conductor in position if the conductor breaks at the point of connection. Protective insulation shall be securely fixed in such a way that it cannot be removed without making the tool unfit for use (e.g. if it is omitted during repair it would be impossible to reassemble the tool in a workable condition) – and so on.

In many situations it is important that fingers, steel rules or even knitting needles shall not be able to touch live or moving parts and a number of probes have been devised to prevent this, including a standard test finger which is hinged and can feel round corners. The subject of detailed design for a particular purpose is discussed in detail in Chapter 4.

1.4 Investigations

Most engineers will, at some time, have to investigate an accident or plant failure. The first requirement is to make sure that one has all the relevant information and that it is correct. Persons who have witnessed a severe accident are often shocked and emotionally disturbed. They

may be quite unable to distinguish between what they have seen and what they think they ought to have seen, or, perhaps what they have imagined when trying to rationalise their confused memories. Some people may also have good reasons for wanting to mislead. The person injured is sometimes less upset and a better witness than the onlookers, for example, a girl who lost several fingers in a guillotine was much calmer the next day than others who saw it happen.

It is also important to remember that the impossible does not happen, and the improbable only happens occasionally. On the other hand, one should always be suspicious of an explanation which comes too readily. With perseverance, the truth can nearly always be found. It is important to examine the debris very carefully after a failure and be very critical of stock wiring diagrams; they frequently have mistakes or refer to the wrong apparatus. Modifications may not have been recorded. For example, after a switch-cubicle explosion which had been attributed to 'a surge', I found, on inquiry, that no system disturbance had been noticed anywhere else. On examining the wreckage, I found that a severed conductor showed clear evidence of a fatigue fracture, not entirely obscured by arcing. The cause of all the trouble, as checked by calculation, was that the conductor had resonated at the supply frequency, work hardened, and fractured while carrying load current.

Having determined how the accident happened, it is important to find out why. Was the equipment suitable for its duty? If an accident occurs because Bill Smith closed the wrong switch, it is important to find out why he did it. Had he quarrelled with his wife before breakfast, in which case you probably cannot do much about it, or was he unfamiliar with the job or the equipment? Were standing instructions vague or ambiguous, was the position of the switch handle misleading, were the circuits confusing, or were the switches inadequately labelled?

Temperament is important in some jobs. A control engineer may have long periods of dull routine punctuated by occasional emergencies when quick and correct decisions are necessary. In such a situation, a man requires enough to do to keep him alert, but not so much that he does not respond instantly when the emergency arises. The UK Medical Research Council Applied Psychology Unit has found that a tired man can usually perform such a job quite as well as a fresh one in normal conditions, particularly if he has had a great deal of experience but may fail to meet an emergency.

1.5 Written reports of accidents

The purpose of an investigation is to ascertain the facts and initiate any necessary action, and a report is usually prepared. If this is muddled or unconvincing the time spent on the investigation has been wasted. A

report should, however, err slightly on the long side rather than be obscure – this is a cardinal sin.

Reports that recommend action by managers or directors not trained in the subject will fail if they are incomprehensible to them. The following points should be borne in mind before writing a report.

1. Think out what your argument is to be and arrange it in a logical sequence so that, as far as possible, each paragraph follows naturally from the one before.
2. Each sentence should carry the argument forward and start in such a way that the reader half expects what is to follow.
3. Be careful with the small change of conjunctions, prepositions, etc. so that the reader is never baulked by anticipating the wrong continuation, e.g. by writing 'and' where 'but' is required; use 'however' or similar words to indicate a change of direction. The correct use of impersonal pronouns is helpful and important.
4. Be careful with punctuation. For example, commas should not be used excessively or mechanically to divide up a sentence like the brackets in an algebraic equation. They can, however, indicate the natural pauses which a good speaker makes to take breath and to help the sense and avoid ambiguity.
5. It is inadvisable to split infinitives and place prepositions unnecessarily at the end of a sentence. Most authorities today agree that the rules on these matters are largely superstitions but your reader may be prejudiced in their favour and ·it is important not to provoke hostile reactions.
6. Avoid long and involved sentences, inversions and wherever possible your own technical jargon which often has a surprisingly limited currency.

Some relevant publications are listed in References 3, 4, 5 and 6. It is useful to have these books but unnecessary to follow them slavishly.

1.6 Accident statistics

If a logical approach is to be made to the subject of accident prevention, it is important to study where accidents occur, and on what apparatus; otherwise time and money will be wasted on misdirected efforts. Statistical analysis is discussed in Chapter 2 but the Tables 1.3 to 1.5 show a typical survey taken over one year.

1.7 Conclusion

The art of electrical accident prevention has been founded primarily on the investigation of accidents by professionally qualified engineers.

Table 1.3 ANALYSIS OF REPORTABLE ELECTRICAL ACCIDENTS BY
LOCATION IN ONE YEAR

Premises	Fatal	Total
Electricity supply	10	101
Factories	19	485
Building operations	5	86
Works of engineering construction	6	18
Onboard ship in dock	1	10
Docks and wharves etc.	–	4
Warehouses	–	2
Miscellaneous	–	6
Total	41	712

Table 1.4 CONDITIONS LEADING UP TO ACCIDENTS IN ONE YEAR

Cause	Fatal	Total
Failure or lack of earthing	5	91
Testing	5	87
Ignorance, negligence, forgetfulness and inadvertence	24	354
Accidents resulting from fault of persons other than injured person	18	160
Working on live gear deliberately	3	108
Misunderstood instructions or failure of permit-to-work system	1	16

There is overlapping between the numbers shown above where more than one
cause has contributed to a single accident. The actual total number of accidents
and of fatalities is therefore less than the sums of the figures in the columns.

Table 1.5 REPORTED ELECTRICAL ACCIDENTS IN ONE YEAR ANALYSIS
BY a.c. SYSTEMS

Standard systems of supply (a.c.)	Fatal	Total
Normal, low and medium distribution voltages (200–450 V single-phase and 3-phase)	23	484
High-voltage distribution (over 3 kV, but not exceeding 12 kV nominal)	13	68
Main transmission systems		
22 kV (nominal)	–	–
33 kV (nominal)	1	4
66 kV (nominal)	–	1
132 kV (nominal)	–	1
275 kV (nominal)	–	–
Nonstandard alternating voltages	2	49

The science of accident prevention is based on a logical analysis of their reports. Some aspects involve highly technical considerations and this is particularly true of investigations of failures where a correct interpretation of small details such as fracture types or surface corrosion, or possible causes of over-voltages is important. But it is an essentially practical subject and its practice is conditioned both by psychological and financial considerations. It is important to spend money first on the action which will bring the greater and if possible quickest benefit.

In closing I would like to make one further point, pay attention to experts but do not be overawed, the greatest experts are sometimes wrong. Therefore verify references and check formulae from first principles if possible.

References

1. *Memorandum on Electricity Regulations*, Form 928, HMSO
2. BS 923: 1972 'Guide on high-voltage testing techniques', BSI
3. Gowers, Sir Ernest (Ed), *Fowler's Modern English Usage*, O.U.P. (1965)
4. Gowers, Sir Ernest, *Plain Words* (Penguin)
5. Jespenson, Otto, *Essentials of English Grammar*, Allen & Unwin (1943)
6. Casey, G.V., *Mind the Spot*, Cambridge University Press

Chapter 2

Statistical Studies and Reliability

List of symbols used in Chapter 2

n	Number of possible results or observations which are equally likely.
p	$1/n$ = chance of a particular result occurring.
q	Chance of a particular result not occurring.
$p + q$	= 1
$^{m}C_{r}$	A binomial coefficient, now commonly written $\lvert^{m}_{r}\rvert$. $^{m}C_{r}$ represents the number of combinations of m things, r at a time.
σ	Standard deviation = $\sqrt{(pqn)}$ for a binomial distribution and its derivatives, e.g. the Gaussian and Poisson distributions.
σ^{2}	= pqn is the variance

If $\quad z = x/\sigma$ where x is the deviation from the mass x

$$\text{Erf } z = \frac{1}{\sqrt{2\pi}} \int_{0}^{z} e^{-\frac{1}{2}z^{2}} \, dz \text{ is the Gaussian 'error Function'}$$

$P_{r} = e^{-n} \, U^{r}/r$ is Poisson's distribution

χ Test. If F = observed number of observations in cell
$\quad\quad\quad f$ = calculated number of observations in cell

$$\chi^{2} = \Sigma \, \frac{(F - f)^{2}}{f^{2}}$$

If f = probability of a component foiling, and
$\quad r$ = probability of the component not foiling is its 'reliability'
$\quad$ then $r + f = 1$

If F = probability of system failure, and $\Big\}$ calculated from f or r
$\quad R$ = system reliability
$\quad$ then $R + F = 1$

2.1 Introduction

Any serious study of safety engineering will show that statistical methods must frequently be used. The purpose of this chapter is to explain the basic principles and their application, drawing attention to pitfalls as well as possibilities. It does not attempt to deal in detail with the mathematics involved in deriving statistical formulae, which are often sophisticated. For this, reference should be made to the standard treatises.

A very important part of statistical analysis relates to statistical distribution, that is to say the way individual measurements or numbers of occurrences vary from mean value. The best known is the Gaussian distribution which is the basis of the classical theory of errors of observation and applies to many other problems, at least as a first approximation. Another is the Poisson distribution of small numbers of occurrences which applies, for example, to the frequency of accidents.

In real life, distributions vary to a greater or lesser degree from these theoretical distributions and these divergences are used to test hypotheses about accident causation. For example the difference in the observed distribution from a Poisson distribution was used in a classical study of individual accident proneness by Yule and Greenwood. This is described briefly in section 2.7.

In many cases there is no *a priori* reason to expect any particular distribution and, for convenience, a selection is made from a number of sets of arbitrary distributions to see which best describes the results. When distributions are found it is often necessary to estimate the probability of wide divergences from the mean. For example, if a steel mast is to withstand high winds we wish to know the highest probable gust velocity and the minimum strength of critical steel sections. Exact values cannot be predetermined but it is possible to calculate the probability of failure and to design the mast so that the probability of collapse is acceptably low.

This leads to the statistical theory of reliability which is discussed in section 2.18. It may be used to identify those critical danger points in, say, a control system where extra precautions are most necessary.

In acceptance testing for switchgear or intrinsically safe apparatus it is impossible to define conditions for absolute safety, but statistical testing procedures have been adopted to ensure that an accident is exceedingly unlikely. This leads to the problem of calamity hazards where a failure is exceedingly improbable but the consequences may be disastrous, e.g. explosives works blowing up. Again, there are two problems; how low must the probability of a failure be, and where can money be most effectively used to prevent it? Which brings in reliability analysis.

2.2 Fundamental considerations

The essential assumption of statistical mathematics is that if there are n equally possible results of a choice, the chance of any particular one being obtained is $p = 1/n$. This does not prove anything, but is merely one way of defining probability. Similarly if there are n ways of choosing six eggs out of a basket the chance of choosing any particular six is again $1/n$.

This is the basis of mathematical statistics; it is an abstraction like the lines without width and points without area of classical geometry. Whether or not it has any relevance in the real world can only be determined empirically. (In fact since Einstein we know that classical geometry is not strictly 'true'.) However, so as to make some progress we will assume for the moment that in some sense probability works in real life. An accident will occur if *any one* of a number of unrelated conditions holds, whose individual probabilities are $p_1, p_2, p_3 \ldots$, then common sense requires that the total probability is

$$P = p_1 + p_2 + p_3 \qquad (2.1)$$
$$= np; \text{ if they are all equal.}$$

The chance of a particular component of a system failing may be expressed as the probability that it will fail within a specified time. If the failure of any one of a number of components (e.g. limit switch on a crane or lift) could cause danger, the chance of danger arising within that period would be the sum of the individual probabilities. Thus if safety depended on the integrity of 500 similar items and any one could fail once in 10 000 hours, the probability of danger arising within that period would be 500/10 000 or once in 20 hours which would be quite unacceptable. This emphasises the importance of very high standards for the components of large and complex products such as a jet airliner, which may contain thousands of critical components.

These matters are discussed in more detail in section 2.18.

On the other hand, if before a system fails, each of several independent safeguards (e.g. interlocks) must first fail, then the chance of total failure is

$$P = p_1 \times p_2 \times p_3 \ldots. \qquad (2.2)$$
$$= P^n \text{ if they are all equal}$$

Again, if any one may fail in 10 000 hours the chance of danger arising will be only once in 10^{16} hours or less than once in 10^{12} years. This illustrates how quickly danger recedes when safeguards are used 'in parallel'*.

*Strictly it is not legitimate to infer that conditions during one period will be valid over an extended period, but it is useful as an indication of very great improbability.

These two ways of combining probabilities are the basis of all statistical investigations, although their actual manipulation frequently involves sophisticated mathematical procedures.

The complement of this approach is to count the number of times we get a particular result in a sample group of observations and if this is x times in y observations we use $p = x/y$ as an estimate of the probability P for the whole field. Unfortunately, it is quite impossible to prove by any logical process that this is valid, and in fact in some circumstances it is not. Our belief that it is a useful procedure is based on past experience of similar situations. Because the sun rose yesterday and many previous days we assume it will do so tomorrow. Without such assumptions orderly life would be impossible. An important part

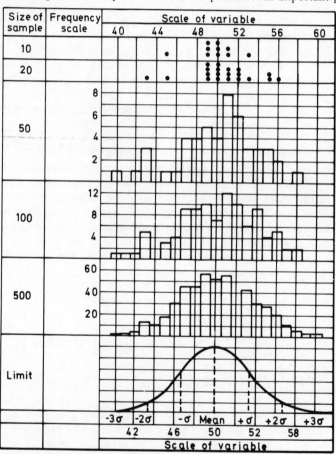

Figure 2.1 Development of frequency diagram from observations (from BS 600)[2]

of statistical theory is to devise appropriate tests of the reliability of such conclusions or such procedures and beliefs.

2.3 The binomial frequency distribution

If, in an extensive series of experiments, the proportion which are successful is p and failures q, so that $p + q = 1$. Then, if we take all possible selections of m tests each, the average number of successes per group will be mp; all individual groups will not have this number but it can be proved that the proportion of the total possible groups which will contain π successes will be

$$P_{\mathrm{r}} = {}^m C_r p^r q^{m-r} \tag{2.3}$$

where ${}^m C_r$ is the coefficient of $p^r q^{m-r}$ in the expansion of $(p + q)^m$; and of course $(p + q)^m = 1^m = 1$. The reason why this arises is that the process of counting r possible p's associated with $(m - r)$ possible q's is exactly the same as determining the coefficients in the expansion. The way sets of individuals (dots) in practice build up into rectangles (histograms) and then into what is virtually a smooth curve is illustrated in Figure 2.1. A binomial distribution is, however, only symmetrical as shown when $p = q = \frac{1}{2}$, otherwise it is asymmetrical stretching further to one side of the mean than the other.

2.4 The Gaussian or normal distribution and the Law of Error

The binomial distribution has only limited uses; more generally p may be very small if m is large. For example, what number of grains of sand will be needed to fill a pint pot? As it would be impracticable to count them and find the numbers in various sample pints we may decide to weigh them and treat the weight x as a virtually continuous variable (which is proportional to the number of grains). It can be proved that the result may be represented in the following way, which is the limit of a binomial distribution for these conditions:

$$\left. \begin{array}{l} \text{If } \sigma = \sqrt{(pqm)} \text{ and } z = x/\sigma \\[2mm] p = \dfrac{1}{\sqrt{2\pi}} \displaystyle\int_0^z e^{-z^2 \frac{2}{2}} \, dz \text{ and } P_{+\infty} = P_{-\infty} = 0.5 \end{array} \right\} \tag{2.4}$$

(Values of P or equivalent integrals are tabulated in books of statistical tables. This is commonly called Efr(z)).

P gives the proportion of times the weight of sand x differs from the mean by not more than σz i.e. the shaded area in Figure 2.2. This figure, unlike a binomial distribution, is always symmetrical.

σ is a very important quantity called the standard deviation and is, in electrical engineering terminology, the r.m.s. value of z. It must

(usually) be determined by measurement because p, q and m are unknown or cannot easily be calculated. σ^2 is also very important and is called the variance.

Although this is in principle the simplest algebraical way of deriving the Gaussian distribution, with the size of the grains all equal, the same curve is obtained if the size varies provided the proportions of the different sizes remain constant in all samples, and the grains are all very small compared with the total volume. It can in fact be shown that not only does the normal distribution represent such comparatively simple cases, but that it can be deduced on much more general grounds and is, subject to certain easily satisfied conditions, the limiting form of distribution for a large number of diverse problems.

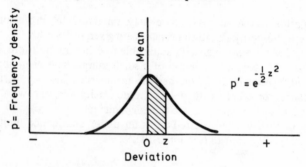

Figure 2.2 *The Gaussian or normal distribution*

Now it is clear that if the error in any measurement is the sum of a large number of small constituent errors which occur with fairly regular frequencies, the distribution of total errors will often follow the Gaussian distribution, and for this reason it has been called the 'law of error'. But it must be pointed out that not all errors or deviations follow this law. The proof of the pudding is in the eating.

It will be worthwhile devoting a little time to the matter as an example of the type of critical examination to which all *a priori* distributions should be subjected. In the first place, considered as a 'law of error' it rests on the assumption that the magnitude of the error is independent of the magnitude of the observation, which is not necessarily true and may be quite untrue. Again, when it is possible to make two related but different measurements such as the diameter and volume of drops of liquid, then if the distribution of one set of measurements is normal the distribution of the other cannot possibly be normal. They are in fact connected by the equation

$$p_y = p_x \frac{d\psi}{dx} \qquad (2.5)$$

where the two measures x and y satisfy the relation $y = \psi(x)$.

2.5 Range

Only very rarely will a result deviate from the mean by more than 3σ, or lie outside a range of 6σ. For a normal, Gaussian distribution, of 20 results the chance of all lying within a range of

$$
\begin{array}{llll}
1.55\ \sigma\ \text{is approx}\ 1 & \text{in } 10\ 000 \\
2.0\ \ \sigma\ \text{is approx}\ 2.3\ \text{in} & 1000 \\
2.5\ \ \sigma\ \text{is approx}\ 3\ \ \text{in} & 100 \\
3.0\ \ \sigma\ \text{is approx}\ 1.5\ \text{in} & 100 \\
3.5\ \ \sigma\ \text{is approx}\ 3.9\ \text{in} & 10 \\
4.0\ \ \sigma\ \text{is approx}\ 6.6\ \text{in} & 10
\end{array}
$$

This matter has been very completely tabulated by Professor E.S. Pearson in Biometrika[3] and some values are given in BS 600.

These figures are not strictly applicable to other distributions. But if the mean and standard deviation of a small sample have been obtained they will give an idea of whether the mean value has any particular significance, or whether they lack homogeneity or vary with some unconsidered parameter (e.g. are time dependent).

This matter will be referred to again under confidence limits and statistical quality control (section 2.14).

2.6 Alternative presentations of frequency distributions

In Figure 2.3(a) the total area of the curve is unity so that the two areas separated by any value such as X represent the probabilities of an outcome greater or less than X. The height of the curve at X is therefore equal to dp/dx and is called the frequency density. This is the conventional (mathematical) way of expressing the matter. This could, for example, be applied to the distribution of fatal electric shocks at varying currents, but it is not always the most useful way of illustrating the problem. Figure 2.3(b) represents the same problem but with the curve scaled to give the relative probability that a particular current X would be fatal. In this case the area of the curve will not be unity.

In practice, although both curves indicate the most dangerous current, very often neither is directly useful because the shock voltage is the only thing we know after an accident and the resistance of the body at the time of the accident cannot be known. What we need to know is the minimum fatal voltage in practice and the range. Neither of these has a precise value and we must base our estimates on past experience, or if we have sufficient data we can define the limits as, say, the voltages

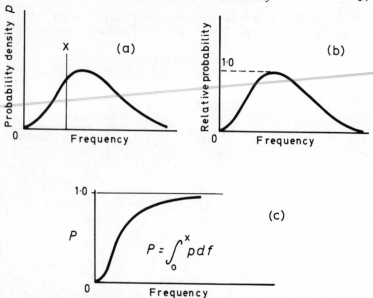

Figure 2.3 Alternative presentations of frequency distributions

between which 999 out of 1000 will occur. This is used to calculate confidence limits (section 2.14).

If Figure 2.3(a) is integrated, we obtain a curve such as that in Figure 2.3(c) sometimes called the cumulative distribution. If the distribution is normal (section 2.4), then the mathematical function it represents is known as the Error function or Erf(x) because of its relation to errors of observation, under certain limiting conditions.

2.7 The Poisson distribution or law of small numbers

If in a binomial distribution p is very small and n large so that $q = 1$ very nearly, then $\sigma^2 = pqn = pn$, which is a small finite number and is in fact the arithmetic mean. It can be proved fairly easily that the probability of a value r recurring is then

$$P_r = e^{-u}u^r/r!$$

(2.6)

where $u = np$ (the mean)

so that if we know the mean, pn, we can both plot the distribution and easily obtain the stand deviation σ, but p and n are usually unobtainable independently.

This distribution gives an adequate description for many accident distributions, for example Rissik[4] gives the information in Table 2.1.

Table 2.1
ACCIDENTS REPORTED IN A LARGE ELECTRICAL SUPPLY
UNDERTAKING

	No. of days on which X accidents occurred	
No. of accidents in a day X	*Actual*	*Calculated*
0	32	31
1	63	64
2	76	69
3	45	48
4	19	26
5	11	11
6	5	4
7	3	1
8	1	1

The calculated values were based solely on the simple fact that 540 accidents occurred in 255 days giving a mean of 2.12 per day. The first thing that one notices is the accuracy of the calculation based on so small an amount of information; it is a characteristic of this distribution that it is reproduced quite well by comparatively small samples. But it is perhaps more important to note that the number of accidents both actual and expected varied from 0 to 8. The standard deviation σ was:

$$\sqrt{2.12} = 1.45$$

Extreme Range/standard deviation $= 8/1.45 = 5.5$

and less than 1 in 60 fell outside the range 0–3.65.

The following are the numbers of fatal electrical accidents in each of 18 years which occurred in factories in Great Britain:
31, 31, 33, 30, 43, 24, 38, 34, 38, 40, 33, 40, 40, 32, 38, 34, 41, 41.

This is far too small a sample to provide a distribution, but the mean is 31.6 the estimated standard deviation 5.6 and the range is 19 which is 3.4σ and is therefore reasonable, so that even the very low value 24 in one particular year has, by itself, no great significance.

The conclusion to be drawn from the above is that when the numbers are small, proportionately large fluctuations have very little meaning, and in themselves do not give much support to conclusions based on other considerations.

2.8 Lack of homogeneity

It is important here to consider the effect of lack of homogeneity in a statistical population, in particular it can be proved that if we take samples by drawing examples, one each from *n* different populations with different values of *p*, then, rather surprisingly, the standard deviation (and variance) will be less than if all were drawn from one homogenous population.

On the other hand, if we drew samples of *n* members from each of the populations in turn, the standard deviation would be increased.

The significance of this is that if we are studying the accident frequency of a mixed population the spread will be less than would be expected by averaging the frequency of the different components, whereas if conditions change over a period of time the spread will be greater than expected from the average values. These two results will work in opposite directions with a mixed population over a period of years.

The following results illustrate the first of these effects. Greenwood and Yule[5] examined the accident records over 5 weeks to 647 women working on HE shells and compared the actual results with those which would arise from a Poisson distribution with the same mean. They were significantly different. (The determination of the significance is discussed in outline below.) They then re-calculated the distribution on the assumption that the accident proneness of the women varied (in a rather complicated way for which there was some evidence). The results are given in Table 2.2 and Figure 2.4.

Table 2.2
ACCIDENTS TO 647 WOMEN IN 5 WEEKS

No. of accidents	Observed frequency	Poisson distribution	Corrected calculation
0	447	406	442
1	132	189	140
2	42	45	45
3	21	7	14
4	3	1	5
5 & over	2	0.1	2

There is substantial support for assuming that different women were not equally accident prone (Figure 2.4). Though the results are compatible with Greenwood and Yule's hypothesis, they do not, however, prove it, there could have been alternative hypotheses. An example of variation with time in a different context is given in Figure 2.6.

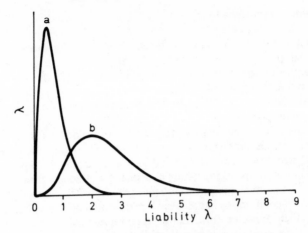

Figure 2.4   *Deduced distributions of liability*
(a) women who had had no accidents
(b) women who had had accidents (Industrial Fatigue
Research Board, Report No 4, Table 9[6])

Another point which may be made here is that in most statistical investigations it is assumed that one selection does not alter those that follow, but if, say, a batch of hospital patients receive a particular treatment and are returned to the ward, statistically the characteristics of the whole may well be no longer homogenous and will have more diversity than before, so that a second 'random' selection would have different probabilities to the first.

2.9 Samples and sampling

Statistical theory suggests solutions to practical problems, which can be empirically verified (or not), or probabilities and distributions are estimated empirically where no *a priori* values are available. In such cases plausible mathematical representations may be obtained, and sometimes by careful analysis, e.g. by studying the difference between the variance of different samples – a possible explanation can be deduced which can be tested by further experiment or observation.

It is generally assumed that a fair sample will be obtained if we make a *random choice* from a *homogenous field*. Many attempts have been made to define these words but have always run into logical paradox and difficulty. If a choice is truly random it cannot be defined. Von Mises discusses homogeneity at great length but in the end his advice is, in effect, be sure you shuffle the pack thoroughly. Ignorance and the 'principle of indifference' do not ensure the result.

Another approach is to deal with the matter operationally. If two methods of selection produce samples which are statistically indistinguishable they may be said to be 'prime' to each other. If a set of results arising by selection or observation are prime to all known methods of selection (which do not beg the question, i.e. the method of selection must not explicitly favour a particular result), we may describe it as homogeneous — although further knowledge may cause us to change our views.

By statistically indistinguishable we mean that when tested by any of the standard procedures derived from statistical theory, any divergencies are found to be no greater than would be expected to arise 'by pure chance'. (See, for example, R A Fisher's *Statistical Methods for Research Workers*.)

In real life, data are rarely homogeneous (some effects of this on the standard deviation were discussed in section 2.8). It is, then, necessary to take sub-samples from all groups which are known or expected to be biassed in particular directions and combine them in proportion to their numbers. The success of market surveys and opinion polls depends on good judgment in this matter; the methods used can be subtle, for example people may be asked what papers they read and what games they play and thus unknowingly give an indication of what their taste or opinions are likely to be in other directions, or vice versa.

A combined selection of this type is the reverse of random but in selecting the individuals from sub-sets a carefully devised random selection is generally used to counter bias on the part of the operators, in the hope of making the selection prime to unsuspected stratification. This is purely empirical and the proof of the pudding is again in the eating.

The philosophy of R A Fisher's use of Random Numbers and Latin Squares in *The Design of Experiments* is similar to this procedure.

2.10 Extreme values

The safety of systems or structures depends not so much on what happens on the average, but what is the worst load or stress which is likely to arise. An important case is wind loading, it is necessary to estimate the most powerful gust likely to occur. If the distribution curve is well defined it is possible in principle to calculate that a value exceeding v will only occur once in x years but, in practice, the shape of an empirical distribution curve is less well defined at very low probabilities than at the mean or mode. This is partly because there is less data but also because rare causes, which have little effect on the general shape may become important, so that there may be a greater number of

large deviations than would otherwise be expected and on this general shape of the curve gives no information.

Questions which arise from such studies are: with limited resources how improbable must a value be for its probability to be disregarded, or at what point does it become better to spend the money available on preventing more likely hazards? This raises the question of protection against calamity risks where the probability is low but the possible consequences unacceptable. This is discussed in section 2.17 and chapters 10, 11 and 12.

2.11 Empirical distributions

By empirical distributions we mean examples where there are sufficient observations to determine the general shape of a distribution, and one wishes to find a suitable mathematical expression to describe it, especially when there is no convincing *a priori* reason to believe that it should have a particular form.

An associated problem is to determine whether an observed distribution can be satisfactorily described by a particular mathematical expression, whether biased on theory, or observation, or a bit of each. Figure 2.5(a) illustrates that there is no logical necessity for the observations to converge on to any well-defined curve, although experience shows that they usually do so. Figure 2.5(b), (i) shows the results of consecutive tests of the electrical resistance of a sample of aluminium powder. Either (iii) or (iv) would represent the matter equally well (or badly).

Sample No	Variation statistically determinate				Variation not statistically determinate	
	Variation statistically uniform		III. One assignable cause present systematically	IV. Combination of III in groups of 4	V. Complete lack of control	VI. Time change present
	I. Smaller S.D.	II. Larger S.D.				
1 2 3 4 5 6 7 8 9 10 11 12 etc.						

Distribution curves representing variation in batch or population

Figure 2.5(a)   Diagram showing types of variation in samples (from BS 600)

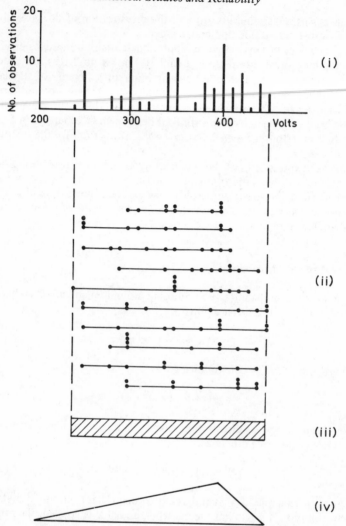

Figure 2.5(b) Results of consecutive tests showing abnormal distribution

Because the results were so erratic and unexpected, powder samples were carefully examined before and after testing and it was found that breakdown occurred by stages, first two, then three and increasing numbers of flakes stuck together forming a conducting chain, which increasing in length as the resistance fell, until a chance jolt separated them; the horizontal rows of (ii), if read from right to left, represent

such chains. (These results arose in the investigation of the possibility of electrostatic ignition of metal powders.)

Presuming, however, that an approximate distribution does emerge on the lines of the samples of 50 and 100 in Figure 2.1(b), the procedure usually adopted is to fit one or other of a number of standard expressions by the so-called Method of Moments. There are various sets of such curves, those most commonly used in the past having in the past been developed by Professor Karl Pearson and expounded in detail by N P Elderton in *Frequency Curves and Correlation* for the Institute of Actuaries.

Other expressions may be obtained by finding the distribution, not of the actual observations but of some function of the observed values, in particular the logarithm, which may be distributed normally or in some known way; the basis of this lies in the relation given in section 2.4 and equation 2.5.

2.12 Weibull's distribution

If $F(x)$ is the probability P of choosing at random an individual having a value X not greater than x, then

$$P(X \leqslant x) = F(x)$$

Transforming $F(x)$ in the following way

$$F(x) = 1 - e^{-\phi(x)} \tag{2.7}$$

we obtain a type of distribution which has proved to have advantages in some engineering problems.

$$Q^n = (1 - p)^n = e^{-n\phi(x)} \tag{2.8}$$

Thus, for example, if a chain has n links and it has been found that the probability of a single link failing when a load x is applied, the probability of the chain surviving load x is Q^n, so that

$$P_n(x) = 1 - e^{m\phi(x)} \tag{2.9}$$

$\phi(x)$ must be a positive, non decreasing function vanishing at a value x which is not necessarily zero. The most simple function satisfying this condition is

$$\phi(x) = (x - x_n)^m / x_0 \tag{2.10}$$

Weibull found that this represented a wide range of engineering and other distributions, e.g. strength of steel, size of fly ash, fibre strength of cotton, fatigue life of steel, etc. Its importance here is that it has been applied to the study of reliability, which is an important aspect of safety (see section 2.18).

2.13 The method of moments and goodness of fit

If the formula for a frequency curve has n independent constants we require n simultaneous equations to determine them. This is, in principle, simple if we know the correct formula and the exact values of the co-ordinates at n points. For example if it were $\log F = a + bx + cx^2$ where F = frequency and x is a measurement, we need only know F for three values of x. In practice, the form of the curve is not usually clearly defined (see Figure 2.5a) and we do not know what the equation should be or the true values of F. We have therefore to try various possible equations, guided by experience, and see which fits the observations best, taking account of *all* relevant observed values of F, which will usually be considerably larger than n in number. The method of moments assumes that the best values of a b and c etc will be those for which the first, second, third (and if necessary 4th and 5th) moments calculated as for the centre of gravity and moments in mechanics, are the same for the calculated and observed values of F.

We shall have to determine, however, which of several possible formulae best represents the result. The most appropriate approach is usually the χ^2 test described below. This is a generalisation of the assumption that the best value of any measurement is the one for which the sum of the squares of the deviations of observed results from the assumed best value, i.e. σ^2, is a minimum. Though this is usually true it is not invariably so, thus for a skew distribution we have three possible best values; the mean, the median, which is the point that divides the observations into equal numbers irrespective of value, and the mode which is the most frequent value and which would be the best value if we wanted, for example, to aim a gun at a target and secure the maximum number of hits. There is an empirical rule

$$\text{mode} = \text{mean} - 3\,(\text{mean} - \text{median}) \qquad (2.11)$$

which is not exact or universally true but is a useful approximation in many cases.

It must be emphasised, however, that the same set of values may be equally well represented by a number of different equations, some fitting better at one point, some at another. Which is the most useful

for a particular purpose is a matter of practical judgment. To test the goodness of fit we may use the χ^2 test. Divide the range of observed values into cells, then

$$\chi^2 = \Sigma \frac{(F-f)^2}{f} \tag{2.12}$$

where F equals the observed number of observations in a cell and f the calculated number. From this value of χ^2 we can find from tables the probability that no worse fit would arise by pure chance, e.g. if the value is 0.05 it means that once in 20 times an equal or worse result should be expected and this is insufficient to reject the equation as incompatible with the observations, but that does not mean that other equations would not fit much better.

If the equation was based on some specific hypothesis as to the nature of the problem the χ^2 test is an indication of its plausibility. Frequently, however, in technical investigations there is no *a priori* expectation of any particular result and the use of a mathematical expression is only a matter of practical convenience. For example, in Figure 2.6 which represents the distribution of momentary loads on a

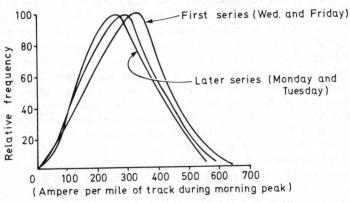

Figure 2.6 Distribution of momentary loads

traction feeder the curve was very clearly defined by the actual observations. It is slightly skew and I tried the formula

$$\log F = a + bx + cx^2 \tag{2.13}$$

and taking 3 values of F obtained a, b and c by solving the three simultaneous equations. The result as judged by eye was almost a perfect fit, and this was adequate justification for the matter in hand which was to find the approximate minimum overload setting of circuit breakers so

as not to cause nuisance operations, and the load factor of the transformer copper losses as a guide to the best design. The curves vary considerably from day to day and great accuracy would be a waste of time (Figure 2.6).

2.14 Statistical monitoring and control (confidence) limits

A great deal has been written about statistical quality control, and it is not therefore necessary to deal with the matter in detail here. Reference can be made to BS 600. The idea is quite simple, some parameter, for example a dimension, is systematically measured on samples of a product and plotted against time as a series of dots or crosses on a chart (Figure 2.7), which should group round the required, or expected,

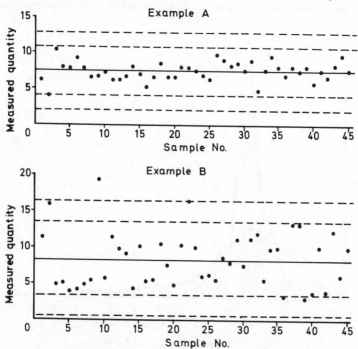

Figure 2.7 Control charts (Dots represent means of consecutive pairs of samples) (BS 600)

value 'like bees round a clothesline', to use the graphic phrase of W.E. Burnand. If parallel to this we draw lines indicating allowable tolerances it will be apparent if the product is and continues to be acceptable. It is

useless, however, to assign tolerances which cannot be met by the methods or equipment in use and therefore they are commonly set on the basis of a statistical analysis to indicate what tolerance will reject, say, 1 in 20 or 1 in 200 items. These are often known as control or confidence limits and for a normal or a Poisson distribution can easily be calculated from published tables.

This maintains a running check on the production standards and also displays any increasing deviations or a gradual drift with time. The first might arise from lack of supervision or care, the latter from wear of machines or tools. If the deviation is persistently to one side it would indicate that either a wrong standard had been set or that there was a systematic error which arose from a mistake in design or planning.

The error distribution may well be asymmetrical; it is then a common practice to average the deviations of a batch of, say, 10 or 20 samples; as the number increases it will, subject to certain qualifications which

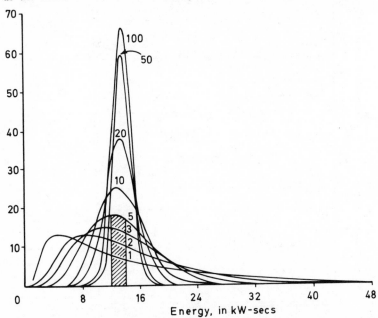

Figure 2.8 Approach to normal distribution
Distribution of the mean energy liberation in groups of n tests. (n having the value indicated against each curve.) Thus the hatched area divided by the total area under curve 5 represents the chance that the mean energy in a group of 5 tests has a value between 12 and 14 kW-sec.

Note. For all curves, the central ordinate of 5 kW-sec, measured by the vertical scale, gives the chance as a percentage that the mean energy in a group of n tests fails in that range.

are usually fulfilled, approach a normal or gaussian distribution. The way this happens is illustrated by an example from a different field in Figure 2.8[9].

It must be noted that although this use of the Gaussian distribution is convenient there is a loss of information. The fact that a distribution is asymmetrical may, if known, lead to improved control limits and/or a better understanding of difficulties in meeting a specification.

2.15 Accident control charts

As Rissik pointed out in 1943, the same type of control chart can be applied to the accident figures of a large company or public corporation. It will be noted that the control lines will not be symmetrically placed, because accident statistics follow a Poisson type of distribution (see section 2.7).

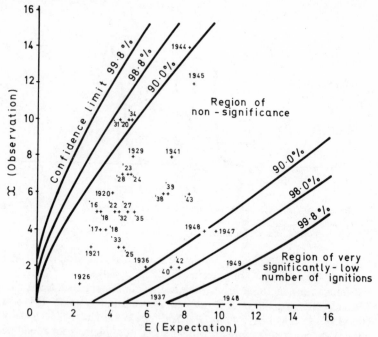

Figure 2.9 Ignitions of firedamp by shotfiring in Great Britain (Safety in Mines Research Report No 7, HMSO)

An alternative method of plotting such figures is to plot actual numbers against expected numbers, based on previous experience. This can for example be used to compare the accident rates in different establishments, see Figure 2.9, based on coal-mining experience.

2.16 Test results

It is well-known that experimental or test results rarely fall exactly on a smooth curve. The requirements of a physicist and an engineer are however sharply opposed on this point. The physicist normally assumes that there is a 'true' value which would be found but for experimental errors or inadequate instruments. The engineer is, or should be, quite as interested in the deviations since in design he wants to know the potential variability of his materials — or imposed loads. Figure 2.10 shows the

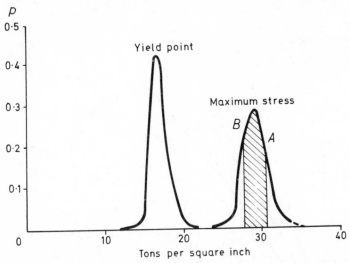

Figure 2.10 Distribution of yield point. The shaded area between A and B represents that proportion whose strength (or elastic limit) was between values A and B. Mathematically, the proportion $P = A\int_A^B p.dt$

distribution of yield point and ultimate strength of 250 samples of steel plate to a single specification taken at the rolling mills.

Similar considerations arise when testing equipment such as inverse-time-characteristic relays and fuses for distribution systems. This was considered in detail by Connor and Smith in 1947. The makers usually publish characteristics represented by a single line, whereas for adequate discrimination between them one needs the characteristics to be plotted as for control charts giving the band within which all examples may be relied on to operate. This is illustrated in Figure 2.11. It is intended that device (*a*) should take over overload protection from (*b*) at point **A** where their nominal characteristics cross but there is an area of uncertainty between B and C (shaded) where, owing to manufacturing

tolerances, it is uncertain which device will operate first. It will be noted that the differences in both time of operation and current represented by B–C are much greater than would be expected from the apparent width of the control limits. This may be very important to operating engineers and seriously affect the safety of equipment.

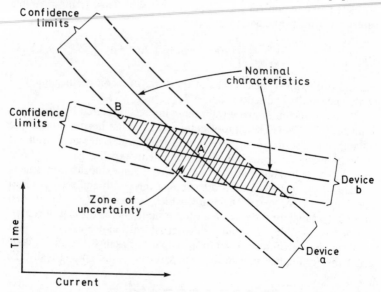

Figure 2.11 Uncertainty in discrimination

There are some proving procedures in which the results vary considerably from test to test. Examples of these are the type testing of circuit breakers where a number of tests are made to determine what is the worst performance likely to arise, and the rating of flameproof and intrinsically safe electrical equipment. An early detailed study of such matters was by Widmore, Whitney and Bruce in 1927 on the number of tests necessary to establish the rupturing capacity of circuit breakers[9], from which Figure 2.8 was taken. The problem here is that developmental and proving tests take a long time and are very expensive, and worse still the results are very variable. One conclusion was expressed as follows: if a circuit breaker survived without undue distress

| 5 tests at 10 000 kVA it was safe for operation at 1200 kVA |
| 10 ,, ,, ,, ,, ,, ,, ,, 2000 ,, |
| 50 ,, ,, ,, ,, ,, ,, ,, 5000 ,, |

but to make 50 or more acceptance tests would have been quite impracticable. Fortunately the operation of circuit breakers is now better

understood and by one means or another the variability of performance
has been drastically reduced, but design and performance are still far
from an exact science. A later paper on this matter was by Steel and
Swift-Hook in 1970 in which they studied the correlation of probability
of failure with recovery voltage, current, recovery-transient frequency,
air blast pressure contact travel, and arc duration. They found tolerable
agreement with formulae of the type

$$P = \frac{1}{2} \left[\; Ef \quad \log h(x/\bar{x})/2\sigma \; + 1 \; \right] \tag{2.14}$$

but did not express their results in the form of a control diagram with
confidence limits as discussed above[10].

A paper which is in many ways more interesting is that by Bruce and
Johnson[11] in 1948. Flameproof equipment, for use in situations where
inflammable gas or vapour/air mixtures may arise, is designed so that if
an ignition should occur inside the equipment, expelled gases will be so
cooled or de-activated that they will not ignite an explosive atmosphere
outside; this is achieved by proportioning the width and separation of
the flanges of the metal enclosure, which is commonly made of cast
iron. Unfortunately there is no clear cut limit, but as the gap or separation is decreased the probability of external ignition decreases.

When the percentage of ignitions is plotted against the gap (Figure
2.12) the results had previously been assumed to be on a straight line

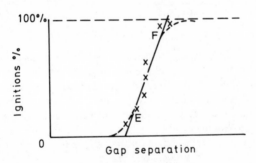

Figure 2.12 Hypothetical example

EF. Each cross represents the mean of a number of experiments, say,
25. Unlike circuit breaker testing the time and cost of tests is not
prohibitive, and a moderately large number can be made; the figure
would represent about 175 determinations.

It had been assumed that the straight line will cut the axis of zero at
the safe value. Several problems arise, however. How many tests must
be carried out for each value of the gap to obtain a fair result; what line

best represents the results; do the mean values really lie on a straight line or should it be slightly curved? Bruce and Johnson conclude that the true curve is slightly S-shaped as indicated by the dotted line.

The problem is best represented by a control type diagram, such as Figure 2.13. What happens at the 100 % limit is unimportant. It is the

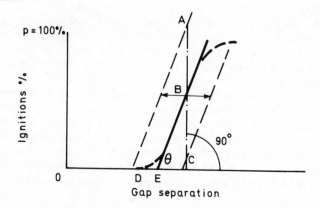

Figure 2.13 Typical control type diagram

position of point D which matters. The important values are D, E and the width of the confidence limits. The spread of the gap values is proportional to $\cos \theta$ and that of the probability p is proportional to $\sin \theta$. Bruce and Johnson assess a gap width such that the probability* of an ignition is $1/10^6$, which is called the statistical safe gap. Because, however, there are still secondary effects which laboratory tests do not cover, a factor of safety is allowed on the statistical gap in BS 229:1957.

Thus we have for example for pentane for 1-inch wide flanges:

Mean safe gap based on a straight line (E)	0.0411 in	Bruce and Johnson
Statistical gap based on a straight line	0.037 in	
Mean safe gap based on a curved line (D)	0.0392 in	
Statistical gap based on a curved line	0.034 in	
Maximum permissible gap for 1 in flanges	0.016 in	(BS 229)

There is thus a factor of safety a bit larger than 2 which leads us to the next section.

---

*Nominal, because although that is the calculation basis, the data do not warrant any degree of accuracy. It is merely a way of saying it is very improbable.

2.17 Factor of safety and calamity hazards

Engineers who design new types of equipment are nearly always working in advance of established theory and in addition, the operation of some equipment is essentially statistical, for example circuit breakers and flameproof enclosures discussed above. Further, the engineers are often only partially informed of the loads or stresses which their designs will need to meet. An architect or structural engineer can only cater for the worst probable wind loading and may have hardly any reliable information at all on the floor loading; even when he is given the weight and type of equipment to be installed initially, he can have no idea what will happen in the future.

For all these reasons, having taken account of all available information, when making calculations they add a factor of safety — which, rather unfortunately, is often referred to as the factor of ignorance. I say unfortunately because this has suggested that as improved methods of calculation are developed and the nature and variability of materials is better understood and improved, the factor of safety can be materially reduced or eliminated. It is worth while looking into this matter more carefully.

Many engineers will know the story of the young draughtsman who calculated, quite accurately, that a small strut should be 3/8 in diameter. The chief draughtsman held up a pencil, looked at it for a short time and said, make it ¾ in. Now this is not just a funny story (or you may think, a chestnut). It embodies the whole theory of factor of safety. Early engineers and stonemasons found that members of a certain size tended to survive, although they made no or few calculations. When 18th century mathematicians began to develop the mathematical theory of elasticity — later simplified into strength of materials and theory of structures — it was found that considerably smaller sections should have survived, but they had not. On the other hand, there was a rough proportionality between calculation and experience and this was crystallised into the 'factor of safety'. Improvements in calculation may lead the young draughtsman to conclude that 11/32 in would be more accurate. But ¾ in will still be the correct size, because experience shows that this is needed to give that extra strength required as a safeguard against those hazards which are essentially unknowable. The law of universal cussedness states that in the long run, if anything can happen, it will.

This leads me to my second point. There are some hazards which are very improbable, but the effects of which are so devastating that provision must be made for them. In actuarial terms, we must consider not only the probability but the expectation which is the product of the

probability and the result, which may be large when the probability is low.

An example was the sinking of the Titanic with the loss of many lives. No one had taken account of the chance of running into an iceberg. Today aircraft carry 300 passengers, soon it will be 500 and 1000 is forecast. But even the best aircraft are occasionally lost, with the lives of all aboard, through unexpected causes. Are we justified in putting 1000 lives at risk and what will be the political repercussions when all are killed in one accident? Another aspect of this matter is that with smaller aircraft when the inevitable failure occurs designs can be modified and weaknesses eliminated with only a few injuries or deaths. With very large aircraft many people may be killed before a weakness comes to light.

To come to more immediate matters, explosions of inflammable gases and vapours may be devestating. Such explosions can be, and almost certainly have been, caused 'by an electric spark'. The electrical equipment may be made very safe, but the best precautions fail occasionally and the proper precaution is to remove or control the hazard. For example, are we justified in concentrating large volumes of potentially explosive material in the middle of large works or in congested urban areas?

2.18 Reliability and planned maintenance

Reliability of components depends on the specification and procedures to ensure not only that the requirements set down are relevant to the use and adequate, but that they are met by sample proving or acceptance tests and are maintained in manufacture over a long period for standard parts. To ensure this either the purchaser or an independent certifying authority should carry out regular inspection of manufacturers' works and test random samples.

Within a works organisation inspection and statistical quality control as set out in BS 9000 is desirable, for critical components at least. Under the BSI Kitemark scheme the instutition carries out routine inspection and test for compliance with a number of standard specifications for small equipment. The engineering insurance companies supervise acceptance tests for materials, etc (e.g. boiler plate) for many of their clients as do the Crown Agents.

The relation between component and system reliability has been explored mathematically by a number of writers and was dealt with briefly in sections 2.2 and 2.10. For more detailed information reference should be made to publications such as those listed in the biblio-

graphy at the end of this chapter. The Review by Wesloski, Low and Noltinck of the Central Electricity Research Laboratories in IEE Reviews gives a list of 104 books and papers.

The mortality of components is generally represented by a curve such as that in Figure 2.14. There are a number of early failures caused

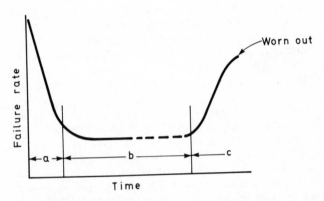

Figure 2.14 The 'bath-tub' curve
(a) 'Burn out period' during which failures caused by defects (b) Useful life during which small number of random failures occur (c) End of useful life

for the most part by fortuitous mistakes, bad material, etc in the initial period, after which there is a long period with sporadic fortuitous failures during normal working and a rise in numbers at the end due to wear and tear which marks the end of useful life.

In section 2.2 above, the manner in which safety increases as interlocks are duplicated was demonstrated. Similar reasoning may be applied to reliability, which for the present purpose can be defined as the probability of a dangerous condition *not* occurring. Thus, if f is the probability of a component failing and r is the probability of it *not* failing, $f + r = 1$.

If the failure of *any one* of n items in a system will lead to danger the reliability is the probability of all of them *not* failing, i.e.

$$R_n = r_1 \times r_2 \times r_3 \ldots r_n = \prod_{0-n} r_n \tag{2.15}$$

$= r^n$ if all r's are equal, which is unlikely.

This resembles the failure of a chain if one link breaks.

On the other hand, if there are *n* safety devices such that all must

fail before danger arises, then the reliability is the probability that not all will fail, i.e.

$$R_n = 1 - F_n = \text{either } 1 - \prod_{0-n} f_n \text{ (or 1} \quad f^n)\qquad(2.16)$$

It is a matter of convenience whether R_n is expressed in terms of f or r.

Our definition of R_n is narrower than the usual definition of reliability since it considers only failures which would lead to danger, but it is my experience that good engineering is usually safe engineering and vice-versa, so that the two aspects of reliability rarely conflict and can be considered together.

In practice we usually have some items the failure of any of which would lead to danger, unless duplicated or otherwise protected, while others are only dangerous in combination, such as two earth faults establishing a sneak circuit. We can also extend the meaning of failure to include any foreseeable circumstances which either alone or in combination are potentially dangerous. A combination of the two conditions discussed above can be expressed as a logical function such as

'If a and b but not c then . . .'

and R_n calculated step by step in the manner described above.

Such proportions can be represented diagrammatically as flow charts as shown in Figure 2.15, where a closed switch represents a healthy item and an open switch a failure. For this example

$$R = r_1 r_2 r_3 (1 - f_4 f_5 f_6) r_7 \qquad (2.17)$$

which may be called its reliability function. In some cases, however, we would also have the condition that an alarm is sounded if any one of 4, 5 and 6 fails.

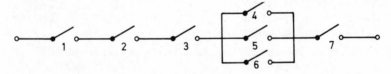

Figure 2.15

In safety studies we are primarily interested in the condition of the system at any time, so that we need to know the momentary value of R, whereas in many other reliability studies it is necessary to know the Mean Time Before Failure which requires a more elaborate calculation; but MTBF has a direct bearing on planned maintenance which has a direct bearing on safety.

If we wish to know the manner in which safety deteriorates with

time, r_n, f_n, etc and therefore R will be functions of time instead of constants. In these circumstances it is an advantage if r_n or f_n can be expressed as $re^{f(t)}$ or $fe^{\phi t}$ and this is an advantage of Weibull's distribution (see section 2.12).

Because the reliability of systems depends on both the components and on the assembly of the whole, it is important to study them carefully. In Chapter 9, a method of simulating control systems on a plug board developed by K Goodman is described. With this it is possible to check the results of mal-operation and the failure of components singly and in combination more effectively than by examining drawings and wiring diagrams.

It was suggested above that failure could be interpreted as any circumstance which could lead to danger directly or in combination. Many fortuitous circumstances cannot be treated statistically because they are unforeseeable, but it is best to adopt the rule that if anything can happen it will, in the long run, and therefore foreseeable contingencies should be treated as certainties, i.e. $r = 0$ or $f = 1$, unless there is back up protection.

Some systems, however, are necessarily very complicated and it becomes impossible to foresee all possibilities and guard against them, and *critical operations must be monitored at the actual point of danger* and effective safeguards devised. Some processes, however, are very fast or take a long time to stop and every effort should be made to predict dangerous conditions automatically before they develop and initiate action to ensure safety.

The statistical approach to reliability is most likely to be of value in showing up which matters which should receive first attention and it and similar studies are useful in comparing different means of protection or maintenance procedures. The mere writing down of the reliability function may indicate the sensitive points.

A simple example of a statistical argument arose when a works engineer decided to check the condition of portable electric tools by taking weekly random samples of 25 % (out of about 500). With this procedure at any time, in round numbers

average number not tested during the previous week would be 375

,,	,,	,,	,,	2 weeks	,,	280
,,	,,	,,	,,	3	,,	210
,,	,,	,,	,,	4	,,	160
,,	,,	,,	,,	2 months	,,	50
,,	,,	,,	,,	3	,,	16
,,	,,	,,	,,	4	,,	5

But this is not the whole story: it is possible on some occasions for the actual number untested to be 2 or 3 times the average at the lower end of the table. For example, in 50 weekly samples we might expect the

number of tools not tested for during the previous 4 months or more to be

0 or 1 on 2 occasions		6 on 7	,,
2 on 4	,,	7 on 5	,,
3 on 7	,,	8 on 4	,,
4 on 8	,,	9 on 2	,,
5 on 8	,,	over 9 on 3	,,

so that there is a substantial probability of a few tools going for long periods unexamined or tested and therefore becoming dangerous (not counting ones hidden away in lockers and tool boxes), whereas if 25 % had been tested each week in a regular sequence, as determined by a register, no tool would go for more than 4 weeks untested.

References

1(a) Bowley, L.V., *Elements of statistics*, P.S. King & Sons (1926). Part 2 gives a very readable account of the derivation of statistical formulae

(b) Kendal, M.T., *The advanced theory of statistics* (2 Volumes), Criffen. This gives a comprehensive account of the mathematics. Because of its length it is easier to follow than some more recent books

(c) Whittaker, E.T. and Robinson, E., *The calculus of observations*, Blackie & Sons. This gives perhaps the most complete study of numerical computation.

(d) Wetherburn, C.E., *A first course in mathematical statistics*, Cambridge University Press.

(e) Aitken, A.C., *Statistical mathematics*, Oliver & Boyd

(f) Moroney, M.J., *Facts from figures*, Penguin books. This gives a wide ranging and very readable account of the uses of statistical theory. There is also a host of recent books from which to choose, mostly directed to particular audiences.

2. BS 600: 1935 *Application of statistical methods to industrial standardisation and quality control*, British Standards Institution

3. Pearson, E.S., 'The probability integral of the range in samples, etc' *Biometrika* **XXXII** Part III and IV (1942)

4. Rissik, H., 'Accident statistics and probability theory', *World Power* (Feb 1943)

5. Greenwood and Yule, *Journal Roy. Statistical Soc.* 83.255 (1920) Quoted in 1(b) above (Vol 1) and in *Industrial Fatigue Research Board Reports* 4 and 28, HMSO.

6. *Report No. 4.* Industrial Fatigue Research Board, HMSO

7. Von Mises, R., *Probability, statistics and truth*, William Hodge (1939)

8(a) Fisher, R.A., *Statistical methods for research workers*, Oliver & Boyd

(b) Fisher, R.A., *Design of experiments*, Oliver & Boyd

9. Wedmore, E.B., Whitney, W.R. and Bruce, C.E.R., '...study of the number of tests required to establish rupturing capacity etc', *Jour IEE* **65** (1927)

10. Steel, S.G. and Swift-Hock, D.T., 'Statistics of circuit breaker performance', *Proc IEE* **117** (1970)

11. Bruce, C.E.R. and Johnson, N.L., 'A statistical method of assessing safety, gaps between flanges of flameproof apparatus', *Jour IEE* **95** Part II (1948)

12. Wesoloski, Low, T.A. and Noltingk, B.E., 'Quantitative aspects of reliability in process control systems', *IEE Review* **119** N8R (1972)

13. Weibull, W., 'A statistical distribution of wide application', *Jour of Applied Mechanics* (Sept 1951)

Chapter 3

The Nature of Electrical Injuries

This chapter reviews the various forms of injury, in particular electric shock, which may result directly from the use of electricity in industry. Mechanical accidents arising from the misfunction of control equipment are not considered since they present no special features to distinguish them from those caused in other ways.

This account is primarily based on personal experience in investigating between 1000 and 1500 electrical accidents of which about one in ten was fatal.

3.1 Types of injury

Electrical injuries are of three main types: electric shock, burns, and falls caused by electric shock. There is a fourth category of very temporary discomfort or incapacity which is not serious, but very painful while it lasts. This is conjunctivitis (or arc eye) which may be associated with shock and burn accidents but is in practice largely confined to electric arc welding.

Other injuries resulting from fires and explosions connected with oil-immersed switchgear, control gear, and transformers could strictly be termed 'electrical injuries' but these are outside the scope of this chapter.

3.2 Electric shock

Serious electric shock is almost entirely associated with alternating currents and is rare when low or medium voltage direct currents are concerned. This is recognised by the Factory Electricity Regulations.

40

Shock is not, however, a single phenomenon but is a general term for the excitation or disturbance of the function of nerves or muscles caused by the passage of an electric current. It is usually painful but is not necessarily associated with actual damage to the tissues of the body. The most common feature is more or less severe stabbing and numbing pain at the points of entry and exit and sometimes along the path of the current through the body. This is frequently accompanied by involuntary contraction of muscles associated with the path of the current (or whose nerves are along that path) which may be painful and torn muscles have been reported.

As a direct result of a moderately severe shock a man may grip and be unable to release a conductor or a tool, or if he has touched a live conductor without grasping it the powerful muscles of his back and legs may contract violently so that he involuntarily springs backwards.

Another possible result of muscular contraction is that the muscles of the chest, diaphragm and glottis may contract strongly and thus prevent breathing, and this might be dangerous and lead to death by suffocation if the victim had also grasped a live conductor (including the unearthed case of defective apparatus) and could not leave go. Death may also follow the arrest of breathing by current passing through the respiratory control centres of the central nervous system; but death is probably most frequently caused by direct interference with the action of the heart. This is usually attributed to ventricular fibrillation. Fortunately the literature, which is extensive, has been usefully summarised and discussed in papers by Francois in 1955[1], by Lee in 1965[2], by Dalziel and Lee[3], by Friesleben and Fitzgerald in 1968[4]. It is possible to obtain a good idea of the trend of research and opinion from these sources.

3.2.1 Hold on current and permissible leakage

Some experimental results on the 'hold on current' are given below and attempts have been made to apply these results to the design of electrical equipment. Once a person is 'held', not only will the experience be very painful but it may eventually prove fatal, particularly because a high resistance fault may easily develop into a more serious failure. There are, however, difficulties in this interpretation and application.

Around 1945, H.M. Electrical Inspectors of Factories became concerned about the dangers from work on live mains-fed radio and television equipment in repair shops. They introduced the concept of earth-free areas in association with safety isolating transformers and sensitive earth leakage protection. It was found possible to design relays which

would operate for leakage currents below the threshold of feeling for these very special conditions. I personally recommended a tripping current *of not more than* one or two mA. (Low values may sometimes be impracticable for stability. A harmless 'tingle' could be regarded as a warning and not necessarily a disadvantage.)

For general use, however, very sensitive leakage protection becomes impracticable because it may be less than the natural leakage, including capacitive-current of the system, and cause excessive nuisance tripping. The values given above are, however, useful in discussing safe leakage current from double insulated equipment and from such conductors as television aerials which are not covered with insulating material.

Because of the difficulty over sensitivity and stability, British and other European practice has tended to rely on voltage limitation for portable equipment rather than leakage protection. Some UK nationalised industries write this into their terms of contract with manufacturers.

3.2.2 Ventricular fibrillation

It is generally believed that the great majority of fatal electrical accidents are caused by ventricular fibrillation of the heart, which prevents its acting as an effective pump, and death follows quickly as a result of lack of oxygen supply to the brain. This is discussed in more detail below.

Experimental work is too dangerous for human beings to be used as subjects, but a great deal of experimental work has been carried out on animals (Figure 3.1) which has been summarised by Dalziel and Lee. It may be concluded that for men the lower (½ % confidence) limit is given by $I = 116/\sqrt{T}$ ma for shocks between 5 and 8 ms and this can conveniently be written $(1160/\sqrt{N})$ ma or $(1.16/\sqrt{N})$ amps r.m.s. where N is the number of half cycles at 50 Hz[6], but the basis of such conclusions is not very secure.

3.2.3 Limitations of experimental results (subjective effects)

The above result needs to be slightly qualified on two grounds. In the first place a 1/200 risk of death is too high to be acceptable when account is taken of the many thousands of painful electric shocks which certainly occur every year in the U.K.

There is an essential difficulty in extrapolating from animal to human subjects, as the latter are considerably affected by subjective

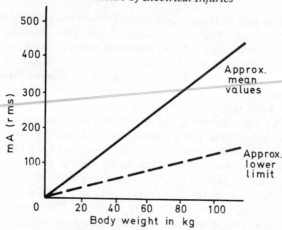

Figure 3.1 Threshold of fibrillation (based on Dalziel)

considerations. For example, it is often believed in the electrical profession that the effect of an electric shock is much greater when either the shock is unexpected or the person is abnormally afraid of electricity. The subjective effect is certainly true of the hold-on current as is shown by the following incidents.

Two cases are known where experienced engineers received electric shocks and were 'held'. They were released when they saw the switch opened, but in each case it was the wrong switch. On another occasion a man collapsed after touching a live conductor, and was detained in hospital for some time. It was later proved that the conductor was in fact dead. In another accident a man received a 25 V shock from a portable lamp, but thought it was at 240 V. He collapsed, was revived, and after collapsing twice more was sent to hospital and detained. The casualty officer accepted the evidence that he had had a severe shock but was very puzzled and stated later that he had strongly suspected that the symptoms were hysterical (although very real to the patient).

For legal purposes it is often important to establish whether a man has in fact died from an electric shock. The difficulty of applying experimental results is illustrated by the following examples.

A man collapsed and died after touching a 12 V dc conductor on a motor car; it is possible that he received an inductive (break circuit) 'kick', but though painful it certainly should not be fatal as the duration would ber very short. No other cause of death could, however, be found at the post mortem examination. Many men receive more severe shocks from sparking plugs, but they are not killed.

I have investigated at least three portable tool fatalities where in each case the man undoubtedly died after an electric shock, but no

fault could at first be found when the tool was tested after the accident. There were clearly intermittent faults of very brief duration. In one case a minute burn on the brush insulation was later found with a low power microscope. In another there was clear evidence that a single copper strand from a rather worn brush tail had slipped through and momentarily shorted between the commutator and ball race. In the latter instance, there were a few very small copper globules, some flattened by the rotation of the race. It was clear that the strand had blown like a very small fuse on short-circuit and that the cut-off effect would almost certainly have limited the duration of the fault to well below one-half cycle (1/100 s). Thus, on the basis of the formulae given previously, the fatal nominal r.m.s. current should almost certainly have not been less than 1.16 A, corresponding at 240 V to a loop resistance including the man's body of about 200 ohms, which is incredibly low for dry surroundings. An explanation might be that he was alone at the time of the accident and was very frightened by an unexpected severe shock, with no immediate help available. It seems probable, therefore, that in some cases at least, rapid 'heart failure' following an electric shock results from other causes than electrically-induced fibrillation.

3.2.4 Body resistance

The resistance of the human body from hand to hand or hand to foot is very variable and depends on the area of contact and whether the hands (or feet) are dry, moist, or wet. Figures from 1000 to 10000 ohms are quoted. Kerwan[5] suggests that RV^k is constant, which Francois[1] accepts, provided k is substantially below unity. The effect of this is indicated by Table 3.1. The figures are tentative but indicate that the increase in safety is much more than proportional to the reduction of voltage, which is in accord with experience and very important. It will be noted, however, that minimum detectable current at 12½ V is in line with the advice given above, and with Tables 3.2 and 3.3.

Table 3.1
CALCULATED ELECTRICAL CHARACTERISTICS OF HUMAN BODY AT 50 Hz IN DRY CONDITIONS

V (volts)	12½	31.3	62.5	125	250	500	1000	2000
R (ohms)	16500	11000	6240	3530	2000	1130	640	362
$I = V/R$ (ma)	0.8	2.84	10	352	125	443	1560	5540

These figures are based on $RV^k = C$ where $K = 1/1.2 = 0.83$ and C has been chosen to give $R = 2000$ at 250 V. The figures are of qualitative rather than quantitative significance.

Table 3.2a PHYSIOLOGICAL SENSATIONS WITH a.c. 50 Hz
Current path: hand-body-hand, over r.m.s. value of current in mA. (According to Friesleben and Fitzgerald)

Physiological sensations		Percentage of test subjects		
		5%	50%	95%
Current just perceptible in palms	at mA	0.7	1.2	1.7
Slight prickle in palms as if hands had become numb	at mA	1.0	2.0	3.0
Prickle also perceptible in the wrists	at mA	1.5	2.5	3.5
Slight vibrating of hands, pressure in wrists	at mA	2.0	3.2	4.4
Slight spasm in the forearm as if wrists would be squeezed	at mA	2.5	4.0	5.5
Slight spasm in upper arm	at mA	3.2	5.2	7.2
Hands become stiff and clenched. Letting go live parts is still possible, slight pain is already caused	at mA	4.2	6.2	8.2
Spasm in upper arm, hands become heavy and numb, prickle all over arm surface	at mA	4.3	6.6	8.9
General spasm of arm muscles up to the shoulders, letting go of live parts just about possible (let-go current)	at mA	7.0	11.0	15.0

Table 3.2b PHYSIOLOGICAL SENSATIONS AT d.c.
Current path: hand-body-hand current in mA. (According to Friesleben and Fitzgerald)

Physiological sensations		Percentage of test subjects		
		5%	50%	95%
Slight prickle in palms and finger tips	at mA	6	7	8
Feeling of warmth and increased prickle in palms, slight pressure in wrists	at mA	10	12	15
Pressure increasing to shooting pain, developing in wrists and palms	at mA	18	21	25
Prickle in forearm, pressure in wrists, shooting pain in hands, increased feeling of warmth	at mA	25	27	30
Increased pain caused by pressure in wrists, prickle reaching up to elbow	at mA	30	32	35
Acute pain caused by pressure in wrists, shooting pain in hands	at mA	30	35	40

3.2.5 The limits of safety

In practical investigations the shock current can neither be predetermined, nor discovered after an accident, particularly after a fatality. The limiting 'hold on' and fatal milliamperes do not help very much, although they are the only suitable units for research work where results are required which are reproducable and comparable between different workers and laboratories. The exception to this rule is their use for deciding on permissible leakage currents in special cases (see section 3.2.1 above). We must therefore base our recommendations on supply voltages, which *are* known, and Table 3.3 is in accordance with experience and practice. There can be no exact determination of these limits but they are given as a guide.

J.G. Wallis repeating earlier tests by D.A. Picken on a number of people found that the *hold-on current* varied from 17 A to 12 A – 14 to 20 V and body resistance (impedance) of 1700 to 2350 Ω. Francois[1] at the Burton Manor conference in 1954 said 'Everything leads us to believe that methods such as those of Schafer, Jellenek, Nielson or Emerson also involve massage of the heart . . . . and that they have a definite effect on the blood reflux in the large vessels owing to the compression of the latter . . . '.

Table 3.3
APPROXIMATE THRESHOLD SHOCK VOLTAGES AT 50 Hz ac

A	Minimum threshold of feeling	10 to 12 r.m.s.
B	Minimum threshold of pain	15 r.m.s.
C	Minimum threshold of severe pain	20 r.m.s.
D	Minimum threshold hold on volts	20 to 25 r.m.s.
E	Minimum threshold of death	40 to 50 r.m.s.
F	Range for fibrillation	50 or 60 to 2000 r.m.s.

Notes: A It is possible for some people to detect whether a pair of conductors is alive at 10–12 V by touching them lightly with the fingertips. If the figure for $k = 0.83$ in Table 3.1 are accepted, this corresponds to a current of approx. 0.7 mA and is in agreement with Table 3.2a.

B,C,D My late colleague D.A. Picken regularly demonstrated these effects when lecturing. But the determination of these limits depends on the individual as the human body can react in many different ways in given circumstances.

E Based on experience over many years and in agreement with British and European regulations and specifications.

F Based on a paper by Francois, Friesleben and Fitzgerald give limits of 80 to 3000 mA. But 80 mA expressed in terms of volts is much too high if Kerwan's formula and table are approximately correct.

3.2.6 The heart considered as a control system

The heart may be considered as a non-linear system with two external control loops and one internal loop (at least). In 1932 in *Chemical*

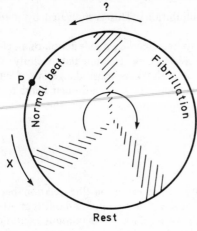

Figure 3.2

wave transmission in nerve, A.V. Hill explained the operation of nerves in electro-chemical terms and A.F. Huxley has compared them to cables in which impulses are propagated without attenuation or distortion; and Professor Power has described the heart's action from this point of view in a very interesting article[6] in *Electronics and Power* which summarises a great deal of published information.

The ventricles or pumping chambers have three characteristic states, rest − normal beat − fibrillation which may be represented diagramatically by Figure 3.1 in which the condition is represented by a point P on the circumference of a circle. If the heart is at rest, healthy, and has an adequate supply of oxygen, it will automatically pass into the 'normal' mode. If it is then violently disturbed, as by an electric current, it *may* pass to the fibrillating mode, but will not necessarily do so. Once fibrillating it will, it is usually believed, continue in this mode until death occurs, but further shocks or manipulation can bring it to rest and if this is not delayed it will recommence beating normally.

There is thus a tendency for P to move round the circle in a clockwise direction. The fact that passage from one mode to another is a matter of relative probabilities is indicated by diffuse (shaded) boundaries between the modes.

The other chambers of the heart (the auricle or atria) can also fibrillate, which may result from certain diseases or damage, but this is not so serious and it sometimes reverts to a normal beat without outside interference.

When beating normally the speed of the heart is subject to the supervisory control of the autonomic nervous system. This is effected by two sub-systems, the sympathetic and the para-sympathetic, which work in opposition. If this balancing act is disturbed the heart may be brought

to rest or nearly so (vagal inhibition). This is indicated by a reverse arrow at X.

Ventricular fibrillation may be caused by direct action of an electric current passing through the heart or in its immediate vicinity; this is supported by experience that a shock is most dangerous if current passes from one hand to the opposite foot, although hand to hand shocks are also dangerous, while shocks between the feet are not thought likely to be fatal.

3.2.7 Effect of frequency

Figure 3.3 shows the results of research on the relation between frequency and electric shock[14]. A.V. Hill first pointed out that 50 or 60 hertz was almost exactly right to produce the maximum excitation of a nerve ending but that the nerves could not respond to substantially higher frequencies. Previously to this the comparative safety of radio frequencies had been attributed incorrectly to 'skin effect' as in ordinary conductors. Radio frequency burns may, however, be serious.

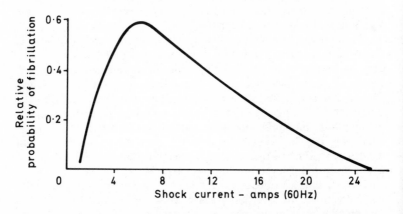

Figure 3.3 Relation between frequency and electric shock based on normal experiments. Such curves are of qualitative rather than quantitative value

Two fatal accidents however threw some doubt on the complete safety of very high frequencies; both were associated with radio frequency plastic heating and moulding. In one case it was believed that a pulsed high frequency voltage may have had the same effect as an alternating current at the pulse frequency. In the other case there was a slight doubt about the source of the current and as the man received

a burn over the heart there was a suggestion that there might have been some thermal or other local reaction.

Finally, if a man's body is in a high frequency field, it will be heated up like any other conductor. It is not known whether there are other effects, but the Post Office issued a pamphlet on 'Safety Precautions Relating to Intense Radio-Frequency Radiation' in 1960.

3.2.8 Shocks involving the head

Dr. Phyllis Croft (MRCVS), on behalf of the University Federation for Animal Welfare, investigated this matter in relation to electric stunning in slaughter houses by shocks through the head and *avoiding the heart* at the Neuropsychiatric Research Unit, Cardiff, and her papers are of considerable interest[7]. We are primarily interested to prevent the *operators* being killed. The main precautions are voltage limitation where possible, and the use of unearthed circuits supplied from safety isolating transformers. Where high voltages are used, as for poultry stunning, because for the insulation of the scaly skin on the legs, the job should be done in an interlocked enclosure. Particulars may be obtained from HM Factory Inspectorate.

In industry, shocks involving the head are rare. I can remember only one fatality: an electrician was found dead behind a switchboard and it was clear that he had touched a live conductor with his forehead while stooping down, quite possibly both his brain and his heart were directly involved. It will have been noted above, however, that a shock involving the head may temporarily arrest breathing.

Controlled experiments were made on rabbits and compared with observations on larger animals in slaughter houses and electro-convulsive therapy patients in hospital.

3.2.9 Respiratory arrest

It appears from Croft's work and experience of electro-convulsive therapy that, in the absence of severe damage to the nervous system, which is rare, respiratory arrest from a shock involving only the head is unlikely to persist, unless presumably the shock has lasted long enough to cause a dangerous reduction of oxygen in the blood (anoxaemia). However, as head shocks are very infrequent, this is not of great importance. It has been my experience that unconsciousness and/or death almost invariably arise from a hand to hand or hand to foot shock, both of which cause a potential gradient in the vicinity of the heart.

Lee states that an experimental shock current of one ampere between

the forelimbs of a small animal does not cause sustained respiratory arrest but that on the other hand, persons unconscious from electric shock sometimes show signs of asphyxia such as turning blue (cyanosis). My personal impression after reading or listening to probably 100 to 150 pathologists' reports on fatalities is, however, that this is uncommon.

3.2.10 Experience of artificial respiration

Data suitable for statistical analysis is not easily found but figures given in *Electrical Accidents*[8] over a period of nine years show that artificial respiration was successful in 47 % of the 400 occasions on which it was known to have been applied. If, however, we consider only the 201 cases in which it was continued for over one hour, unless the patient

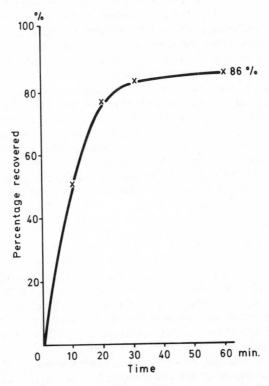

Figure 3.4 Probability of recovery when AR is continued for at least 60 min unless recovery occurs earlier

recovered earlier (and no-one recovered after an hour), we get the following result:

112 or 56 % recovered in 10 minutes or less
153 or 77 % recovered in 20 minutes or less
165 or 83 % recovered in 30 minutes or less
172 or 86 % recovered in 60 minutes or less

Thus, only 29 or 14 % died when AR was continued for over an hour and 86 % recovered — not 47% as above.

Estimates of the lapse of time under emotionally disturbing conditions cannot be very accurate but although the sample 201 is not very large a clear pattern emerges (Figure 3.4). The 112 who recovered in 10 minutes or less were possibly not deeply unconscious, some may not have been much more than dazed, but 60 recovered after substantial periods of AR and it is highly unlikely that they would all have done so without AR.

There is another aspect of these figures. Probably all the persons who recovered in these years (1942–50) will have been treated by the Schafer (or similar mechanical) methods. There is thus a clear suggestion that the Schafer method still has value, particularly when only one person can administer artificial respiration and external heart massage is impracticable or unsuccessful.

3.2.11 How important is artificial respiration?

We have arrived at an apparent contradiction. Pathological and other evidence suggests that only a minority of deaths are due to asphyxia, and therefore the rate of success of artificial respiration should not be large; but it has in fact been 86 % successful when persisted with. If the majority of deaths are in fact due to ventricular fibrillation which is commonly thought to be quick and irreversible in the absence of heart massage or other forms of defibrillation, AR should be ineffective, which it was not.

Therefore if only a minority of these 201 men suffered respiratory failure, how did a substantial number survive 20 or 30 minutes and longer with ventricular fibrillation? It seems clear that at some point during this time fibrillation, if it occurred, must have stopped. Best and Taylor state that ventricular fibrillation *may* do so, though infrequently. The figures given above imply that either they were not fibrillating at all, or that when AR is applied, by the Schafer method at least, the chance of reversion to a normal beat is quite good (This is represented by a reverse arrow at the top of Figure 3.2). If the fibrillation did not cease very quickly the brain must have received sufficient blood (and therefore oxygen) to keep it alive and it has been suggested that

this could be maintained by the rhythmic pressure on the heart and the rectifying action of the valves.

It is clear that there is a considerable gap in our knowledge here and further research is necessary but this is not easy, because it is not possible to experiment on human beings (although the ancients did so on slaves and prisoners). Experiments on animals are of limited applicability because, apart from other considerations, experience suggests that the results with men may depend to an extent on psychological factors.

There is one other observation which may possibly explain a few anomolous cases. Helen Taussig describes the case of a man struck by lightning who received first-, second- and third-degree burns on his face, shoulder and great toe. He was taken to hospital in a semi-conscious condition, breathing slowly with normal blood pressure but irregular pulse. An electro-cardiogram confirmed a diagnosis of *auricular* fibrillation. Could auricular fibrillation contribute to some puzzling cases of near electrocution in industry?

When a man appears to be unconscious and not breathing artificial respiration is (or should be) immediately commenced. Both works 'first aiders' and electricians are normally instructed to do this since a medical diagnosis is very rarely quickly available and prompt action is essential; often such accidents occur in remote or very inaccessible situations. AR should be continued until regular breathing is resumed or for a minimum of one hour. (When breathing recommences artificial respiration should continue until it is regular and the operator should then stand by to give further help if it falters.) Our knowledge of the effectiveness of AR in practice is necessarily based almost entirely on records of such action, since nothing better is available.

3.2.12 Conclusion

A man who has been revived by artificial respiration may remain in a critical condition for some time and need further respiration before he is safe. It is important for ambulance attendants not to remove a man who is just breathing but still not fully conscious unless they can continue artificial respiration until they arrive at hospital; this is usually possible nowadays. In the past, cases have occurred where a man in this condition has been given no treatment during the journey and pronounced dead after a cursory examination on arrival and sent direct to the mortuary.

My general advice to anyone faced with an apparent case of electrocution is to immediately start artificial respiration by any system you know, and do not bother about the relative efficiencies. Do not let anyone stop you in less than one hour; you cannot injure a dead man.

I feel however that we should make quite certain that mouth to

mouth respiration is at least as effective as Schafer for cases of 'electrocution' (as distinct from gassing and drowning) before it is entirely abandoned. This also is a subject for further research.

Dr. P. Croft[7] wrote a series of reports on the electrocution and electric stunning of animals for the University Federation for Animal Welfare from whom details may be obtained. Major C.W. Hume, the Secretary General of the UFAW, has written a very interesting pamphlet entitled 'Electrocution – a historical retrospect'.

3.3 Other injuries

3.3.1 Acoustic shock

Accounts of the after-effects of lightning strokes frequently refer to temporary or permanent impairment of hearing and sometimes ruptured eardrums. This is in most cases almost certainly caused by the intense acoustic shock wave sent out when a column of air (the lightning channel) very rapidly expands on being suddenly heated to about 15 000°. The range of danger is probably limited to a few feet. Although an exploding wire such as a small conductor or fuse on short circuit may make a very loud noise, I have never heard of serious acoustic shock in industry.

3.3.2 Arc eye or conjunctivitis

This is a very painful condition resembling pepper in the eyes and develops some hours after looking (even momentarily) at an intense source of ultra-violet light. It may occur when a person is very close to a severe short circuit but it is primarily confined to work with electric welding and potentially to arc furnace workers.

Fortunately 'arc eye' only lasts a short time, not more than a day or two, and although painful leaves no permanent injury. It is in effect a form of sunburn. The only treatment is the application of a soothing lotion. Some lotions have contained small amounts of cocaine; although this appears to be effective in relieving pain, there appears to be some risk of addiction if it is used too liberally.

This complaint is easily prevented by the use of goggles with effective side protection. Welders themselves seldom suffer because they have to use goggles or visors. The usual victims are assistants setting up work who do not use goggles and other persons in the vicinity, for example, crane drivers. Fortunately ordinary clear glass cuts out much of the ultra-violet light and special goggle glass which virtually eliminates it is

only slightly tainted. There is a British Standard covering suitable protective goggles.

Theoretically there is a risk of a form of cataract caused by prolonged exposure to infra-red light which affects persons who work for long periods on glass furnaces etc, but I have never heard of a case among workers on electrical plant.

3.3.3 Fractures and torn muscles

Many textbooks refer to torn muscles and fractured bones caused by violent muscular contraction, but I have had very little personal experience of this. Strains and fractures may also arise from falls following an electric shock, e.g. from a crane or ladder.

There is no special problem about fracture and normal medical attention is all that is required, except that occasionally it may be necessary to make certain that a person has not been 'knocked out' and perhaps seriously injured by a fall when found unconscious. If there is a doubt whether the victim has been stunned or shocked mouth-to-mouth respiration is probably to be preferred. It is a matter of judgment in the particular case.

3.3.4 Burns

A person who has received an electric shock and survives usually suffers no injury unless he has had a severe fall or been burned. Burns are in fact the most serious after-effect of electrical accidents and are the principal danger with direct currents or very low voltages (below about 80 V), whereas shock is the typical injury from low and medium voltage alternating currents, although there may also be severe burning. At extra high voltages shock may not be so important but burns tend to be very severe and may cover a large area of the body. Burns may be of several types, as described below:

(a) *Contact burns* where the patient has touched a live conductor. These may be local and very deep reaching to the bone, or on the other hand very small, just an area of 'white' skin which may be easily overlooked at a post mortem examination. The position of such small burns may be very important in reconstructing an accident and should be recorded.

(b) *Arc burns* may be very extensive, and of any degree, particularly when there has been a high voltage flashover. Provided the patient survives the initial wound and surgical shock, and the surface area involved is not too large, he is likely to make a good recovery as the

injury is sterile. He may, however, be badly scarred or lose a limb. In the 1930's such burns were largely confined to high voltage circuits. More recently, however, increasing power and transformer capacity has sometimes made severe arcing on medium voltage circuits a serious problem. For example, on one occasion where persistent arcing occurred on a medium voltage switchgear panel, the arc and hot gases involved the whole width of the switchboard passageway and several fitters had to escape on their hands and knees.

(c) *Radiation burns from short circuit arcing*. Though this is in effect a severe form of sunburn, it can be disabling. On one occasion, five professional engineers landed up in hospital after a short circuit on an open medium voltage switchboard, one of them was approximately 22 m from the short circuit.

(d) *Vaporised metal*. When an open fuse or small conductor fuses some copper (silver or tin) is vaporised, and at close quarters this may burn or impregnate the face or hands. This is not common, but on one occasion a man's spectacles were 'plated' with copper or copper oxide – which saved his sight.

(e) *Deep burns and necrosis*[10]. Jalenik, Frolicher and other continental experts have stressed the danger of deep burns destroying tissues below the skin where superficially there is only a small injury. They emphasise that electrical burns, and in particular high voltage contact burns, must be taken very seriously and the patient kept under medical supervision. I have, however, known only one clear case of this type of occurrence. A man received a small burn on his leg; it was dressed in hospital and he was sent home. About 8 or 10 days later he developed pneumonia and died. There being no apparent cause, a post mortem examination was made and it was found that the cause was a pulmonary infarction (blood clot lodged in the lung) which was almost certainly a direct result of the electrical injury which was found to be much deeper than it had appeared to be.

(f) *Metal fume fever*. I have personally investigated only one case of metal fume fever caused by inhaling metal or metallic oxide fumes by a welder working in an enclosed space. This is not however an unusual condition and it emphasises the importance of good ventilation. Fortunately, though unpleasant, it was of brief duration. (This case was only referred to me because there was a doubt about the technical circumstances)[12]

(g) *Nervous and psychological side effects*. Over the years I have talked to a substantial number of persons who have received severe electrical shocks. Very few have complained of persistent after effects, either mental disturbance, headaches and nervous pains or debility, where there has been no obvious physical damage. In the few exceptions I have been unable to judge whether their subsequent troubles were a

result of some physical condition or the psychological effect of a frightening experience on a nervous person. Quite possibly I have not by any means always heard a full account of the after effects of the accidents I investigated, but I am fairly confident that the prospect of a full and rapid recovery is usually very good.

For some reason which I do not understand, psychological and functional disability appears to be much more common among people who have been shocked or injured by lightning discharges than following serious shocks in industry.

3.4 Conclusion

Electric shock is not a single simple phenomenon and is not perfectly understood, but practically everyone today is liable to receive a dangerous shock sooner or later since the great majority of electrical accidents occur at the common domestic and commercial electricity supply voltages, i.e. 240 and 450 V respectively. One (reportable) electrical accident in ten is fatal. The following paragraphs deal with accidents and their treatment up to the time the person is seen by the casualty department.

Contrary to some recent statements artificial respiration for persons unconscious after a shock is very effective if continued for at least one hour, in fact there is an 85 % chance of recovery, but there are a number of matters on which further research is desirable. In the past the Schafer method has been very successful. The most important thing, however, is to use whatever method you know and start as quickly as possible. An unconscious man does not need much oxygen to keep him alive and fine points as to the relative efficiency of manipulative methods of artificial respiration are unimportant (this has already been fully discussed earlier in this chapter). The position is quite different for gassing accidents where the maximum ventilation of the lungs is essential.

Electrical burns may be extensive and deep but are sterile and therefore tend to heal quickly and well although they may leave scars; damage to muscles may be serious and amputation may be necessary in very bad cases.

Particular attention should be paid to small burns caused by contact with high voltage conductors because there may be serious deep seated damage (necrosis) which is not visible. Because of this, full information should be provided to casualty departments at hospitals, preferably in writing.

Artificial respiration should be applied until the patient is breathing freely and regularly and must be continued during the journey to

hospital if necessary. The patient should be under close observation for some time after he recovers as there is some danger of a relapse. Full details, preferably in writing, should again be sent to the casualty unit.

Comparatively few people have ever seen a case of electrocution and there is a need for better education at all levels, from the Medical School to the first-aider. Electric shock is quite different from drowning and gassing and the appropriate treatment is not necessarily the same.

References

1. Francois, R.C., 'Recent experimental and clinical studies and present trends in research, occupational safety and health', *Occ. Safety & Health* (1955)
2. Lee, R.W., 'Death from electric shock', *Proc IEE, Vol* 113 (1966)
3. Dalziel, C.F. and Lee, R.W., 'Evaluation of lethal electrical currents', *Trans. IEE Industry and General Applications*, VIGA (April 1968)
4. Friesleben, K.J. and Fitzgerald, B.D., 'Electric shock – assessing the danger', *Electrical Times* (12 Dec 1968)
5. Kerwan, L., 'Measurement of the resistance of earth circuits and of the human body', *Congres Technique National de Securite d'hygiene du Travail* (1950)
6. Power, H.M., 'Marvellous control systems of the heart', *Electronics and Power* (Feb 1975)
7. Croft, Phyllis, G., 'The effects of electrical stimulation of the brain in the perception of pain', *Jour of Mental Science* **XCVIII** 412 (July 1952)
8. HM Factory Inspectorate, *Electrical Accidents*, Published annually, HMSO
9. Fordham-Cooper, W., 'Electrical safety in industry', *IEE Reviews. Trans IEE* 117 (Aug 1970)
10. Jellinek, S. and Froliche. Lectures on electric shock at Conference held at Burton Manor College (1954)
11. Taussig, Helen B., 'Death from lightning and the possibility of living again', *American Scientist* (7 March 1969)
12. *Fume from welding and flame cutting*. Report on the Shipbuilding and Ship-repairing Industry, HMSO (1970)
13. Best, C.H. and Taylor, N.B., *The physiological basis of medical practice*, Bailliere, Tindall & Cox, London (1950)
14. Based on figures quoted by Francois and Dalziel. Original work was carried out by Ferris, Ring and Spence in 1936.

Chapter 4

Detailed Design (Double Insulation)

4.1 Introduction

This study covers the supply of electricity to portable and transportable electrical equipment without the necessity of providing facilities for earthing. It takes account of the recommendations given in BS 2754 'Memorandum on double-insulated and all insulated electrical equipment'. It is associated with Chapters 7 and 8 and is an example of the care necessary in detailed design briefly discussed in Chapter 1.

The use of double-insulation marks a stage in the development of safe design because, where, previously, double-insulation was largely confined to domestic and similar premises, amendments to the *factory* Electricity Regulations now include conditions for the use of double-insulation in industry. Briefly the regulations require that only apparatus which has been tested and approved by a *recognised authority* may be used and that it must be maintained in a safe condition to comply with the conditions for certification.

The certifying authority for domestic types of equipment in Great Britain is the British Electrical Approvals Board and for other apparatus it is usually the British Standards Institution. The phrase 'approved authority' allows for the use of imported apparatus which has been tested elsewhere to substantially the same standard by other internationally recognised authorities. Apparatus for use where a flammable atmosphere may arise must ber certified as safe for that purpose, see Chapter 11.

4.2 General principles

The principles of double insulation are set out in BS 2754 which has been recognised internationally as the clearest statement available of

the requirements of such equipment. It is not itself a specification but is intended to guide committees drawing up specifications and assist manufacturers in carrying them out.

Double insulation is defined as 'a method of insulation by which accessible metal parts are separated from live parts by both functional and protective insulation'. Functional insulation is defined as 'the insulation necessary for the proper functioning of the equipment and for basic protection against electric shock'. Protective insulation is defined as 'an independent insulation provided externally to the functional insulation and functionally insulated parts, in order to ensure protection against electric shock in case of failure of the functional insulation'.

In this context functionally insulated metal parts are not the live conductors covered by functional insulation, but are the inaccessible metal parts, insulated from live parts by functional insulation and are themselves covered by protective insulation. The protective insulation may itself be enclosed in, or support, an outer metal case or other metal fittings which are accessible, i.e. likely to be handled by the user. It is intended that this accessible metal should not be earthed and that safety from electric shock should depend on the fact that, before accessible metal becomes live, two independent layers of insulation must break down. The construction has some resemblance to screened or pliable armoured cable, except that it is not intended that the screening should be earthed.

The principle of double insulation has been accepted as an alternative to earthing by the IEE Regulations for the Electrical Equipment of Buildings, provided, and the importance of the proviso will become apparent, that domestic apparatus had been made and tested in accordance with BS 3456 and certified by the British Electrical Approvals Board (see section 4.1).

4.3 Difficulties of interpretation and construction

4.3.1 Insulation failure

It is clear that if we are to dispense with the earth connection we must be certain that the alternative protection is not less reliable. Earthing has one great advantage in that earth continuity can easily be checked by inspection and simple test equipment. When, however, only the functional insulation has failed in a piece of double-insulated equipment it will still work, no one will get a shock and no fuse will blow, and the appliance will have become in effect a piece of unearthed single-insulated equipment without anyone being aware of the change. The equipment may be used in this condition for a long time, until indeed failure

of the protective insulation causes an accident. It is important therefore, the integrity of both layers of insulation should be checked periodically and independently. This is recognised in BS 2754 which states that it is desirable that means of access for this purpose should be available without dismantling the equipment. Alternatively, it should at least be possible to make separate tests after such dismantling as can, if necessary, be done with simple tools such as a screwdriver.

Where there is a layer of functionally insulated metal between the outer case and the conductors, it is not difficult to arrange these separate tests.

Although it will usually be necessary to remove a cover plate and perhaps insert a probe into a hole in the protective insulation, since functionally insulated parts are by definition inaccessible from the outside a difficulty arises when there is no intervening metal screen. The construction here resembles an unscreened cable with insulation round the individual conductors and an insulating sheath overall, which may have external accessible metal braiding. In such a case the test pressure recommended is 4000 V. There is some doubt, however, whether this is completely satisfactory. If the layers are of different thickness and/or of different dielectric constant they will not be equally stressed and one part may not be adequately tested. There is, moreover, another difficulty. High voltage testing is not directly correlated with conditions of service; its main purpose is to detect gross faults such as pinholes and metallic inclusions in insulation, incorrect assembly or inadequate manufacturing tolerances. While, therefore, it is a useful empirical test for new equipment, it is not necessarily entirely satisfactory as a routine test for equipment in use, or which has been repaired.

4.3.2 Insulation resistance

The more usual test for apparatus in use is to measure the insulation resistance, but this also will only detect certain types of fault and will not distinguish between protective and functional insulation unless they can be tested separately. A further difficulty is that functionally insulated metal does not necessarily form a continuous sheath so that even where the functional and protective insulation can be tested independently, this may entail testing at more than one point, and this necessity may not be evident from the outside. This applies to both pressure and insulation resistance tests and is perhaps most likely to be overlooked by the user.

It is clear that type tests or even routine pressure tests by the manufacturer cannot alone guarantee safety, and reliance must be placed on

manufacturing technique and strict quality control to ensure the safety of appliances as sent out.

4.3.3 Mechanical considerations

In most situations, functional and protective insulation should have somewhat different properties. The former is itself protected from mechanical harm by the construction of the appliance and it is frequently securely fixed or at least supported for the greater part of its surface. The material used may therefore be selected for its ability to prevent leakage and to withstand unavoidably high temperatures. The protective insulation may, however, be exposed to form part of the outer casing of the appliance – or the whole outer casing in the case of all-insulated appliances. It must, therefore, be chosen largely for its mechanical properties.

All specifications, therefore, prescribe fairly severe impact and associated tests, the criterion being not that the apparatus shall continue to work, although that is desirable if it can be achieved, but that its safety shall not be impaired. Particular attention should be given to resistance to abrasion, e.g. by dragging equipment over a rough floor.

4.3.4 Misplaced insulation and loose connections

In the past, it has been known on electrical tools for the terminal screw on the trigger switch to come loose and touch the metal casing of the handle. If this handle were not earthed a serious accident would have been almost inevitable. With double-insulated equipment it is usual for the switch and similar parts to be completely enclosed in a chamber of insulating material which can be a casing or capsule.

When the tool is repaired it is important that this insulation material is not disturbed. To avoid this, regulations state that any such component must be securely fixed in position and cannot be removed intact, or that it shall be of such a size and nature that it cannot well be omitted, e.g. the other parts would obviously not fit together properly without it. This also applies to any barriers, insulating tubes, screens, etc. whether forming part of the functional or protective insulation.

Similarly, the design should be such that no insulation can be bridged or clearances and creepage distances reduced by a loose or detchaed nut or screw, or by a detached or broken lead. A weak point on leads is adjacent to a terminal or a soldered joint, and fracture here may be overcome by supporting or clamping it a short distance away. These precautions should prevent a very high proportion of all the

electrical faults which cause shock accidents and should be applied to earthed as well as double-insulated appliances.

4.3.5 Apertures in casing

Any hole in the outer casing of an electrical appliance is a possible source of trouble. There is, therefore, a general requirement that only those gaps necessary for the proper functioning of the equipment should be permitted. In practice this usually means entry and exit ports for useful air in the case of such appliances as hair dryers, and for cooling air in motor-operated appliances. There is a difference here between earthed and all-insulated and double-insulated appliances. If a screwdriver, or a skewer, is pushed through a hole in the metal case of an earthed appliance it will most probably cause a short-circuit and any shock is likely to be momentary. This does not apply to all-insulated equipment, and with double-insulated equipment the whole case becomes live. Some test is therefore necessary. There has been some confusion on this point but it is generally agreed that the standard test finger is too large to keep children's or some women's fingers from touching live internal metal close to an opening, and that a smaller probe is necessary. BS 2754 requires that such openings 'shall be of such a size and so disposed that the test finger cannot be pushed through them to touch either live metal or functionally insulated parts *or functional insulation* less than 2 mm thick, and as an additional safeguard against contact by a metal implement it should not be possible to touch live parts with a test *pin*'. It should be noted that there is a reason for the double test, the test finger is jointed and can therefore touch parts not in the direct line of sight which the straight pin, representing a screwdriver or a nail file, or even a knitting needle, cannot reach.

4.4 Thickness of insulation and creepage distances

The creepage distance and clearance in electrical equipment is often very small — sometimes not more than a millimetre. These values must be looked at very carefully in double-insulated equipment and minimum values have been laid down. Great difficulty has been experienced, however, in specifying values which are adequate for normal construction but which are not unduly restrictive for special purposes or inadequate in, say, a damp situation such as a stone-floored kitchen on washday with the air full of steam and the walls covered with condensation.

The IEE Wiring Regulations admit double-insulated apparatus in such a situation, and since it is impossible to keep portable apparatus in

a particular room, all apparatus must be suitable for these conditions. The thickness of external insulation is of crucial importance if it is to withstand abrasion, as by dragging across the floor. It must be remembered that the functional insulation may be an air gap between live metal and the protective insulation, so that a very rigid construction is necessary as well as resistance to abrasion as mentioned above.

If values are specified which will not hamper the development of new materials and techniques they may be inadequate for some existing material and apparatus. It is desirable that any testing and certifying authorities should have a certain discretion in dealing with such matters and that they should not feel obliged to approve apparatus about the safety of which they may have doubts.

4.5 Testing and certification

From this account of some, but by no means all, of the problems which arise in connection with double-insulation it should be clear, that, although the underlying principle is easy to understand, it is far from easy to make satisfactory and safe equipment, and that unless the development is under some sort of control a dangerous situation could arise. This type of inspection is carried out by the BSI, BEAB and some recognised overseas authorities, for certain apparatus.

This is particularly important for double-insulated equipment since some faults in manufacture may not be detectable in a completed appliance without destroying it (e.g. it may be necessary to cut a section through a motor armature or a moulding with insets) and the defect may go undetected until a further fault causes an accident. The appropriate safeguard here is carefully supervised manufacturing processes with inspection during manufacture and strict quality control.

In the UK the use of double-insulated equipment is only accepted in industry if it is made in accordance with the appropriate standard specification, supervised and certified by a recognised testing authority and maintained to that standard in use.

Acknowledgement

This chapter is based on material originally published in the *Electrical Times* and is reproduced with the permission of the editor and publishers.

Chapter 5

Mechanical Causes of Electrical Failures

This chapter develops in further detail a point made briefly in Chapter 1, namely that the safety of electrical equipment depends very largely on sound mechanical design. It is a statistical fact that the majority of circuit breaker failures, for example, are mechanical rather than electrical and good design and quality control in manufacturers' works is essential to eliminate this type of defect. The information in this chapter emphasises this (see also Chapter 7).

5.1 Introduction

In the early days of electricity supply, transient conditions were very imperfectly understood and any occurrence resulting in arcing or flashover for which there was no apparent cause was vaguely attributed to a surge. Often more careful investigation strongly suggested and sometimes conclusively proved that the real cause was a mechanical failure of either a conductor or its supports or some similar occurrence.

In this chapter attention is drawn to different ways this can happen by describing a number of failures drawn from a wide field. It must not be assumed however that failure is common. In fact, switchgear and control gear has a good record of reliability, but for this reason, when failure does occur, it is easily attributed to the wrong cause and a false diagnosis may entail a waste of time and money without effecting a cure.

5.2 Simple examples

There are, of course, silly mistakes and small oversights which occasionally cause serious failures or accidents. Typical examples were when a

fitter left some cotton waste in a circuit-breaker tank (which exploded when the waste got in between the contacts on closing), or the failure to provide adequate barriers between adjacent compartments of m.v. switch and control panels, thus allowing bits of wire or washers to fall or poke through.

Attention must also be drawn to the danger from loose joints in conductors which lead to overheating and sparking. The former reduces the electric strength of air and the latter may cause ionisation and both may lead to 'flashover'. An extreme example of a loose joint was found when compound-filled trunking on the supply switchgear of a large machine exploded. It was found that the clamping bolts in a busbar joint had been completely omitted and contact had been maintained for several years by the rigidity of the compound filling, until an overload softened it.

A somewhat similar trouble arises from ineffectively filling with compound h.v. cable boxes, particularly switch, transformer and motor terminal boxes. If voids are left, either moisture may enter or ionisation may occur and lead to a short-circuit. The explosions caused may be violent and at least one man has been killed, but apart from this the number of accidents to persons is small compared with the number of failures. A suggestion has been made that boxes should, as a matter of routine, be examined by gamma ray radiography.

5.3 Mechanical resonance and fatigue fractures

The danger of resonant vibrations in current-carrying conductors is illustrated by a series of arcing faults in an 11 kV oil circuit-breaker cubicle. The first fault extensively damaged the auxiliary equipment, but although there had been widespread arcing and the cubicle doors were blown off the switch itself was only damaged externally. No reason could be found for this failure, no other fault had occurred and there was no other system disturbance.

In the course of repairs extra insulating barriers were inserted at the point where the trouble appeared to have started, but almost exactly a year later a precisely similar failure occurred in the same cubicle and although the barriers had reduced the damage, it was still serious. Again no reason could be found and, with rather less confidence, further barriers were installed. After another year a third similar failure occurred and, on this occasion, during an investigation made immediately after the failure, it was found that the severed end of a ½ in diameter rod conductor was partly melted by arcing but otherwise showed a typical fatigue fracture.

The obvious explanation was that the rod was of such a length that it resonated under the normal, comparatively weak, electromagnetic forces and had become fatigue hardened and broken, which took about a year, and this fracture of a current-carrying conductor was the cause of the trouble. The accurate pre-determination of resonant frequencies is difficult; trouble is therefore best prevented by introducing a degree of mechanical damping.

5.4 Corrosion fatigue and stress corrosion

Without going into metallurgical details it may be said that where metal is stressed in a corrosive situation, e.g. when exposed to a damp or polluted atmosphere, particularly when there are alternating forces, failure may occur at comparatively low stress.

An interesting example of this was found in the investigation of an obscure arcing fault within a tubular porcelain insulator on some 11 kV switchgear. This was another case where arcing had been caused by the fracture of a current-carrying rod, probably associated with vibration similar to that described above, although this could not be proved conclusively.

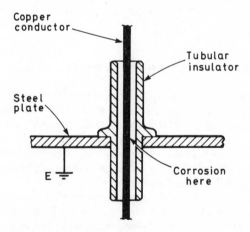

Figure 5.1 Corrosion fatigue failure of conductor

It was noted that there was a green patina or deposit on the rod on either side of the break where the arcing had started. This rod was examined by a metallurgist, who stated that corrosion had been caused by nitric or nitrous fumes and that it appeared to be a case of stress corrosion. When the conductors on the other two phases were

examined these also were found to be corroded at the corresponding point, but had not reached the point of failure.

The matter is illustrated diagrammatically in Figure 5.1 from which it will be seen that the fracture occurred at a place where there was a concentration of electric stress, caused by the proximity of the charged conductor and the sharp edge of the hole through the earthed steel chamber. Ionisation at this point had presumably caused the formation of oxides of nitrogen.

The obvious remedy is to reduce the electric stress which here indirectly caused mechanical failure, without reaching the point of direct electrical breakdown. The corollary is to select, where necessary, materials which resist corrosion. It may be noted here that some atmospheres have a very serious effect on particular metals, for example, brass can become weak or disintegrate by the solution of zinc from the brass in the grain boundaries, but still retain its outward form. Tannic acid in oak will destroy lead sheathing, turning it ultimately into lead carbonate or white lead. Copper is liable to attack by some salts, particularly those of oxidising acids, such as chromic acid.

5.5 Resilience, stress raisers[1] and elastic distortion

The resilience of a structural member may be defined as the amount of energy it can absorb without significant damage. The theory of this is discussed at length in such books as Morley's *Strength of Materials*[2] and Salmon's *Materials and Structures*[3] and there is no need to repeat it here, but it will be convenient to state a few basic results for reference.

The simplest example is of a uniform bar in tension; the impact energy it can absorb per unit volume is $(f^2/2E)$ where f is the stress and E is Young's modulus. But if the cross-section is reduced it will stretch more for an equal load; it will absorb more energy *so long as the elastic limit is not exceeded*. If, however, a bar is of varying section diameter the thinnest part will reach the elastic limit and foul first, absorbing less energy.

The standard illustration of this point is that of a bolt in tension (see Figure 5.2). In (b) the cross-section is reduced to that at the root of the thread and therefore on this simple theory the reduced bolt will absorb more energy than a bolt of full section.

The root of the screw thread is, however, an inherent weakness as it is a point of stress concentration. This applies also to any sharp re-entrant angle reduction of section in a member under stress, particular shock stress or stress reversals. In current jargon these are 'stress raisers' (i.e. design features the result of bad machining which causes a local concentration of stress from which a crack may be propagated).

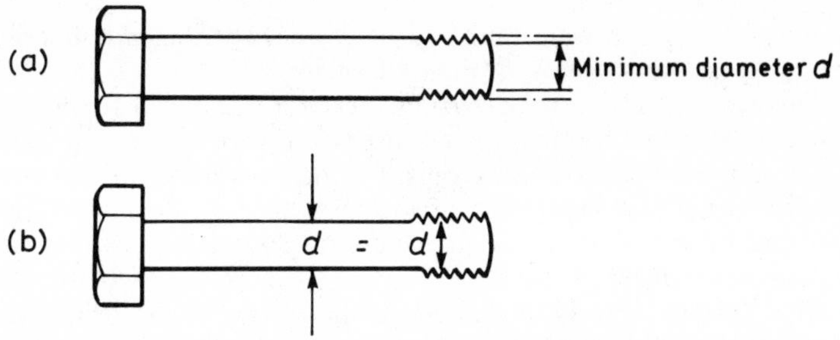

Figure 5.2 Demonstrating the effect of resilience (a) normal bolt; (b) bolt with reduced shank. Bolt (b) will absorb more energy than bolt (a) because it will stretch more. (Energy absorbed is equal to ½ maximum force × elongation.)

The design problem where fluctuating or shock loads are applied is to obtain the maximum resilience without local concentration of stress or unacceptable distortion under load.

5.6 Examples of failure[4]

An example of failure under repeated shock loading is provided by a number of circuit-breakers, made by several leading manufacturers, when used for controlling arc furnaces at about the beginning of the last war. Normally a power supply breaker operates perhaps once a year and rarely more than an average of once or twice a day, but these breakers were operating 10 000 to 20 000 times a year, a duty for which they were not designed and for which they were not suitable. There were a number of explosions and some serious accidents; on several occasions when there had been nothing of that kind, arcing contacts and whole cross-arms were found loose in the bottom of the tank.

When a circuit-breaker closes, a rather heavy mechanism is rapidly accelerated and then suddenly stopped as the contacts meet, and the energy has to be absorbed. In several cases it had been found that fractures occurred at screw threads or at sudden changes of section. Such failures may arise from inadequate radii at the roots of flanges, etc. Needless to say, once it was appreciated that these equipments were subjected not to a single or a small number of shocks, but many thousands, the designs were modified and the trouble disappeared.

Unacceptable elastic distortion may, however, be equally serious. When an oil-circuit breaker clears a heavy fault, flammable gas is generated and a mixture of oil and gas may be expelled at the junction between the tank and the top casing. Experience of investigating a considerable number of failures leads to the conclusion that although

the gas may be ignited in the tank this is not always so and sometimes it is ignited on coming into contact with the outside air. In one instance, a switch lost several gallons of oil. Internally it was virtually undamaged and the spigot tank joint showed no signs of weakening after the event, but an external oil vapour explosion bulged the 14 in walls of a large switch house and let down the concrete roof slab on to the top of the switchgear.

Assuming that under the effect of an internal explosion it is impracticable to prevent some stretch in the bolts, the less the stretch, the less oil will get out and in addition the greater the chance of extinguishing any flame — on the principle of flameproof enclosure. It is also clear that the mere provision of accurately machined flanges and bolts of sufficient strength and resilience is not enough in the design of flameproof enclosures. Possible stretching of bolts or distortion of the flanges must be considered, particularly if large enclosures or light alloys are to be used.

5.7 Brittle fractures[5]

Some metals, e.g. steel, though metallurgically identical, may fracture in two entirely different ways under slightly different conditions. This has been known at least since 1875, but it was not until World War II, when a number of ships broke in half for no obvious reason, that any considerable amount of research was carried out.

It is now well established that some steels which, under normal tensile test conditions exhibit considerable ductility, will, at low temperatures, fail by brittle fractures with no ductile deformation. This is important because the safety of many engineering structures and components depends, in part, on slight plastic flow relieving local and largely fortuitous stress concentrations. This is particularly important if these 'stress raisers', are small, unnoticed cracks. There is some evidence that brittle failure is more likely to occur if a heavy load is applied suddenly, although it may also occur when conditions are stable (Figure 5.3).

Some years ago two flameproof controller enclosures burst, both were in the open or only partially protected from the weather. In the first case the ground was covered with snow, and the second failure occurred in December. In this latter case the cover was blown off, virtually neither damaged nor bent, eighteen bolts having failed in tension. A simple calculation showed that bolts were too small but the manner of failure was surprising and suggested brittleness.

In the first case the cover consisted of a steel plate welded to a thick rim through which the bolts passed, not a very good construction; here

the weld failed all the way round with virtually no permanent set in the steel plate. Both covers were about 20 in. by 30 in. These failures contrast strongly with other occasions when for example the top plate of a transformer has failed, a few bolts have broken and the lid has remained attached but badly buckled.

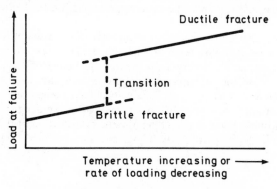

Figure 5.3 Transition from ductile to brittle fracture

It is impossible to be certain now whether either or both of these failures were in fact examples of low temperature brittleness but, when it is remembered, that large switchgear and transformers are often sited in the open, the possibility of low temperature brittleness occurring should be considered. Quite small components may be subject to low temperature brittleness. Impact, notched bar, bending tests or the more recent notched bar impact tensile tests, at varying temperatures are sometimes used to indicate the transition temperature from ductile to brittle fracture (Figure 5.3). Fatigue cracks from alternating or fluctuating stress may act as stress raisers and precipitate brittle failure.

5.8 Fracture of insulators

Mechanical failure of ceramic insulators may displace conductors and cause short-circuits or, alternatively, spread the trouble once it has started. Often one finds only a mass of fused metal and shattered porcelain which cannot be interpreted, but sometimes it is possible to obtain a lead either by examining damaged metal parts as described above or the broken remains of insulators. Ceramics are essentially brittle materials which are strong in compression and comparatively weak in tension and will normally fail in tension with no plastic flow. For example, in pure bending the main fracture will be normal to the bending axis, and when twisted a cylinder will tend to fail at 45° to the axis.

Perfect textbook fractures may be obtained by breaking a stick of blackboard chalk as illustrated in Figure 5.4, which may be compared with those given for, say, cast iron in Morley[2] or Salmon[3]. Fractures illustrated in textbooks, however, are produced under carefully controlled conditions to ensure that the only stress is the one under consideration. Such elaborate simplicity does not occur in real life. A clue

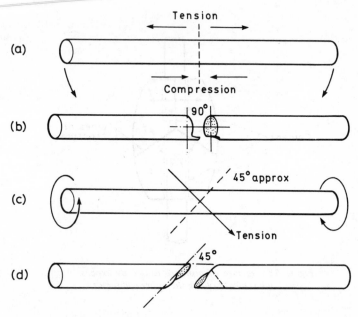

Figure 5.4 Brittle fracture of a rod (a) and (c) showing the application of the stress while (b) and (d) indicate the type of fracture that ensues

as to what has happened may, however, be found by considering the planes of principal stress under conditions of combined stress. The way this leads to a helical brittle fracture for twisting is illustrated in Figure 5.4 (c) and (d).

5.9 Causes of insulator stresses

Stress in insulators can arise in a number of ways. Perhaps the most straightforward is inaccurate alignment of rigid parts which will usually cause a bending moment, and inaccurate or badly aligned bedding, particularly where an insulator is under compression. This causes local stresses near the base. Other causes are electromagnetic forces during

short-circuits or thermal expansion of conductors or the insulators themselves. On occasions, it has appeared probable that a through fault has cracked busbar insulators causing widespread arcing at points remote from the initial source of trouble.

Bursting stresses may arise from the insulator cement swelling, (Figure 5.5), although not much has been heard about this recently. An

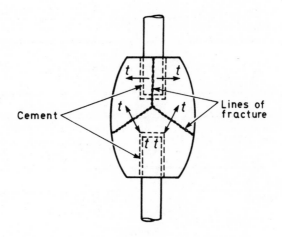

Figure 5.5 Fracture of insulator caused by expansion of the cement holding the upper conductor

interesting example occurred a number of years ago when litharge and glycerine cement exposed to a contaminated atmosphere turned into lead carbonate. This surprising result was checked in two ways, the X-ray spectogram of the cement was compared with that of pure lead carbonate, and a sample of new cement was exposed to flue gases, and checked in the same way with the same result[5].

A bursting stress may arise in a badly designed oil-filled insulator resulting from overload or short-circuit heating. This should not occur, but a new aspect of this problem is found with insulators filled with insulating gases at high pressure to suppress corona and discharges. If they burst, lumps of ceramic may be thrown a considerable distance.

In a thin cylinder under internal pressure, the longitudinal tensile stress is $pd/4t$ where p is the internal pressure, d is the diameter and t the thickness, while the circumferential tensile stress is $pd/2t$, so that it will tend to split axially. This may be upset by mechanical constraints (e.g. end caps) and variations in thickness. For comparatively thick

cylinders the problem is more complex. The tensile hoop stress is proportional to

$$\frac{R_2^2 + R_1^2}{R_2^2 - R_1^2}$$

where R_2 and R_1 are the outer and inner radii and it is greatest at the inner surface. A crack will, therefore, start at the inner surface and be propagated outwards.

An extension of this problem arises when the inner surface of a tubular insulator is suddenly heated. To determine the rate of spread of the heat outwards it would be necessary to solve Fourier's equation

$$\frac{d\theta}{dt} = K \nabla^2 \theta$$

for the particular circumstance.

All that need be said here, however, is that the heated inner part will expand and be under compression, which must be exactly balanced by tension on the outside (Figure 5.6). At first no failure will occur, but

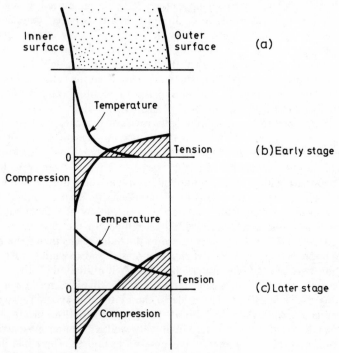

Figure 5.6 Changing temperature and stress distribution in thick cylinder heated internally, (b) early stage (c) later stage

as the heating spreads outwards the internal compression area will increase and the exterior hoop tension will also increase until the outer surface splits (parallel to the axis) and a crack is propagated inwards. An example of this occurred some years ago when several 2 kV cartridge fuses in a substation cleared satisfactorily but split an appreciable time afterwards; they fell to pieces, which caused considerable trouble in replacing them.

It is interesting to note that, in theory at least, if a cylinder is heated internally by gas under pressure the pressure will cause tension on the inner surface which the thermal stress will tend to cancel.

5.10 Cable failures

As an example of a fault which was in fact very different from what it appeared to be, the case of a small cable may be mentioned. It had been subjected to a very heavy through fault and presented an alarming appearance. At approximately every 2 m it had swelled and 'bird-caged' the armour, and the obvious conclusion was that some sort of standing voltage wave had been set up causing it to be punctured at regular, short intervals followed by local short-circuits between conductors.

A little consideration will show that this was highly improbable. It would correspond to a very high-powered oscillation of very short wavelength and high frequency. However caused it would have 'out-magnetroned' the magnetron and would have been the fulfilment of the radar engineer's dreams. For this reason a section was cut out and examined. The lead was bulged and slightly split under the armour and the outer bonding papers were split for several inches, and the core papers were scorched and torn; there was no sign of arcing or any puncture although the paper was badly overheated and discoloured.

The only explanation available seems to be that these were local hot spots, possibly caused by kinking and hardening during laying, or while supported over brackets on some other site. That a considerable temperature was probably reached was shown by the fact that the copper near the end of the cable had melted. In the absence of any electrical fault the most likely cause of the bulge seems then to be that the impregnating oil was partially evaporated and expanded as a gas bubble between the core insulation and the belt insulation.

Another example of a cable fault not being what it seemed occurred during a switch failure. The cable to the switch was apparently punctured and earth current had burned away part of the lead sheath. This explanation did not, however, fit in very well with other evidence. A section of cable was therefore cut out and examined. There had been no insulation failure, but earth current from the frame of the switch had arced over to the cable sheath.

5.11 Conclusion

It is hoped that the examples given have sufficiently illustrated the importance of examining damage carefully before ascribing it to a vague and, therefore convenient, cause. The examples also illustrate the importance, even to electrical engineers, of having a sound knowledge of mechanical design and properties of materials as well as electrical properties. More could be done on these lines in mechanical courses for electrical engineer. When discussing fractures metallurgists and engineers tend to talk different languages and ceramicists and electrical engineers are hardly on speaking terms. Instruction on the appearance of fractures and failures in the untidy way they occur in practice rather than on a testing machine would be valuable[7].

The practical investigation of failures may be compared to clinical medicine and pathology as opposed to the tidier studies of anatomy and physiology.

References

1. The Institution of Mechanical Engineers has published a series of data sheets on 'Stress Concentration Data'
2. Morley, Arthur, *Strength of Materials*, Longmans, Green & Co.
3. Salmon, E.H., *Materials and Structures (Vol 1)*, Longmans, Green & Co. (1931)
4. *Insulating Oil in Relation to Oil Circuit Breaker Failures*, W. Fordham Cooper, *Journal IEE* **90**, p. 11 No. 13 (Feb 1943)
5. R.H. Redmayne, *Private Communication*
6. A concise account of this matter was given in the *New Scientist* for 15 April 1965
7. Official report on the Flixborough Explosion, HMSO (1975)

Chapter 6

Abnormal Circuit Conditions- Over Voltages

Many electrical failures and injuries arise from over currents and excess voltages and their consequences, the cause of which is often obscure. In this chapter which is intended as an aide-memoire rather than a systematic study, a number of related results are collected together for convenience of reference.

In practice it is often unnecessary to make calculations during an accident investigation. In fact reliable numerical data are not usually available after the event, but in tracing the cause of a failure it is of great assistance to keep the physical and mathematical principles clearly in mind as this suggests what to look for and where. On the occasions when accurate calculation is necessary the standard treatises should be consulted[1-6].

Some of the electrical hazards described in this chapter may seem unlikely, but they do occur occasionally and must be correctly diagnosed if time and money is not to be wasted on useless modifications, and the hazards remain.

6.1 Initial conditions

On closing a switch and energising a system of conductors the initial current usually consists of two components, the normal amount I_a determined in phase and magnitude by the circuit and load impedance and a damped dc and/or ac transient I_d, which are initially equal and opposite so that if $\bar{I}_{a_1}$ is the initial (instantaneous) value of $\bar{I}_a$

$$\bar{I}_{a_1} + \bar{I}_d = 0 \qquad (6.1)$$

List of symbols (used in Chapter 6 and 7)

The following convention has been adopted to distinguish clearly between the various aspects of current and voltage.

CURRENT

I	General symbol	
$\overline{I}$	Instantaneous value	
i	Amplitude of a.c. current $I(\cos\theta + j\sin\theta)$ or $I\underline{/\theta}$	These are particular values
I	$\mathcal{J}$Amplitude of a.c. current $\mathcal{J}\mathcal{J}(\cos\theta + j\sin\theta)$ or $\mathcal{J}\underline{/\theta}$	

VOLTAGE

V   General symbol
V, ν, V etc As for current

GENERAL

X_L	Inductive resistance
X_c	Capacitive resistance
f.	Frequency
ω	$2\pi f$
M_{nm}	Mutual induction
L or M_{nn}	Self inductance

ϕ. Magnetic flux. ϕ_L . Leakage flux. ϕ_m . main or mutual flux*

L	Reactance
C	Capacitance
$IRRRV$	Instantaneous rate of rise of recovery voltage
$RRRV$	Average rate of rise of recovery voltage
rms	Root mean square
t	Time
θ	Temperature

$[I^2 t] = \int I^2 dt$. Integrated between limits appropriate to a particular application, e.g.
    Pre-arcing $[I^2 t]$ of a fuse
    Let through $[I^2 t]$ of a fuse or circuit breaker
    Withstand $[I^2 t]$ of a component, e.g. a diode

*Note that ϕ_L and ϕ_M are not the main and leakage fluxes as usually understood but are defined in section 6.3 in a manner convenient for calculating the types of voltage transient (following the convention in ERA publication *Surge Phenomena*)

If a change, such as a fault or switching operation elsewhere suddenly changes the appropriate current

$$\overline{I}_{a_2} + \overline{I}_d = \overline{I}_{a_1} \quad \text{initially}$$

$$\text{or} \qquad \overline{I}_d = -\Delta\overline{I}_a \qquad (6.2)$$

and is again damped. I_a which may be direct or pulsating may be called the 'bridging' transient. Bridging transients are frequently at relatively high frequencies.

6.2 HF transients

High frequency transients have an important bearing on safety, for example they can materially affect the ability of a circuit breaker to operate on short circuit. For a simple circuit the frequency and maximum amplitude are easily calculated. In Figure 6.1(a) an alternating

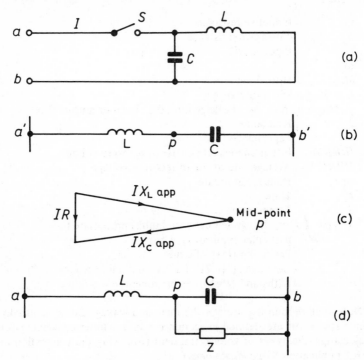

Figure 6.1 Elementary circuits

current I is assumed to be interrupted by switch S without arcing or sparking, and the current is thus suddenly diverted to charge the condenser, thus setting up a high frequency oscillation at frequency f.

The resistance may be neglected because serious trouble will not normally occur unless this is small, otherwise the oscillations would be effectively prevented by damping. The conditions round the closed charging loop, there being no source of energy, give

$$-v = \mathscr{I} \; (X_L - X_C) = 0$$

where v and $\mathscr{I}$ are the amplitudes,

$$X_L = 2\pi fL \quad \text{and} \quad X_C = 1/2\pi fC$$

so that

$$\sqrt{f} = 1/(2\pi\sqrt{(LC)} \quad \text{or} \quad \omega = 1/v(LC) \tag{6.3}$$

At one extreme the (undamped) energy in the circuit is $\tfrac{1}{2}L\,\mathscr{I}^2$ at the other extreme it is $\tfrac{1}{2}Cv^2$ and these must be equal so that

$$v = \mathscr{I} \sqrt{\frac{L}{C}} \tag{6.4}$$

and therefore

$$V = I\sqrt{\frac{L}{C}} \quad \text{r.m.s.}$$

This approach has been adopted in most of the following discussions because it greatly simplifies formulae and brings out the essential features more clearly. Most engineers will be able to assess the changes damping would produce.

6.3 Multiple frequency transients

A network, neglecting damping, has as many modes of oscillation as there are independent closed loops. An important example is an unloaded three-phase transmission line, Figure 6.2(a). The figure is of course simplified and is only intended as an illustration of the principles. It is equivalent to Figure 6.2(b).

There are three loops which may be called loop 1 comprising the two phases a and b, loop 2 comprising b and c and loop 3 comprising

c and a. Then proceeding as before we have three simultaneous equations of the form

$$\varkappa = \mathscr{J}_1(X_{L_1} - X_{C_1}) - \mathscr{J}_2 X_{M_{12}} - \mathscr{J}_3 X_{M_{13}}$$

where $X_{L_1} = 2\pi f L_1, \quad X_{M_{12}} = 2\pi f M_{12}$

$$X_{C_1} = 1/(2\pi f C_1) \quad M_{12} \equiv M_{21}$$

where L is self inductance and M mutual inductance and C capacitance.

From these three simultaneous equations, three frequencies (which are unlikely to be equal) can be calculated in the same manner as for the single-phase case and we are left with three amplitudes and three-phase angles to meet the initial conditions. The highest possible voltage

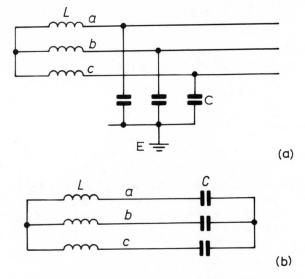

Figure 6.2 *Unloaded three-phase transmission line*

would occur when all the initial energy (usually predominantly inductive, i.e. $\Sigma\frac{1}{2}LI^2$) has been transferred to one capacitance, and the maximum possible current occurs with all the energy associated with one inductance.

If there were in fact no damping, these values would always eventually occur unless the frequencies were exactly equal – which is virtually impossible – because over a period the energy is concentrated

first in one member, then another. This can be illustrated by analogy with two pendulums of different periods supported from the same flexible support. If one is set swinging it will progressively come to rest as the other takes over and then vice versa. Damping will, however, tend to suppress this action so that the values thus calculated will usually not be reached.

6.4 Three-phase systems

The most important instance of multiple circuits arises with three-phase systems, and it is not always easy to visualise the manner in which the operation of a circuit breaker, for example, affects the instantaneous currents in the three phases. A modification of the method of trilinear co-ordinates is useful for this purpose[7]. In Figure 6.3 the three lines *Oa*,

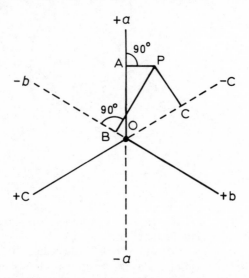

Figure 6.3 Three-phase systems

Ob, and *Oc* intersect at *O* and the projections of a point *P* on the three lines are *OA*, *OB* and *OC*, then, algebraically

$$OA + OB + OC = 0 \qquad (6.5)$$

to an appropriate set of scales. If the lines are at 120° to each other, as drawn, the three scales are equal. The proof of this special case is quite simple.

The point *P* may therefore represent the values of the instantaneous currents in three conductors whose sum is zero, and in particular in a cable or overhead line of a three-phase system when there is no overall leakage or out of balance current. Similarly it would represent the voltages to neutral in a balanced three-phase system. If *OP* rotates about *O*

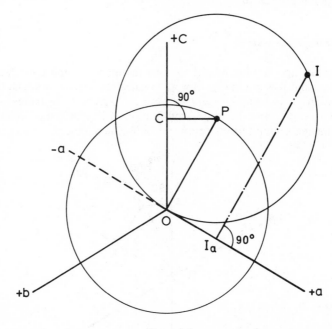

Figure 6.4

as in Figure 6.4, then *P* will represent the current variations with time in both phase and magnitude for all three phases.

Further, if *I* rotates uniformly about *P* with radius *OP* (i.e. passing through point *O*) it will represent three currents whose sum is always zero, and each of which is the sum of a constant and an alternating component, and all three currents are zero at point *O* when the constant and alternating components are all equal and opposite. This is the initial state of a symmetrical three-phase short circuit on an unloaded system and, neglecting decrement, the initial parts of the three current 'oscillograms' can easily be drawn (Figure 6.5). This is useful, for example, in studying the effect of 'point or wave of application' and asymmetry in short circuit tests on three phase circuit breakers. On phase *a* for example the current is nearly symmetrical, whereas on phases *b* and *c* the currents are very asymmetrical.

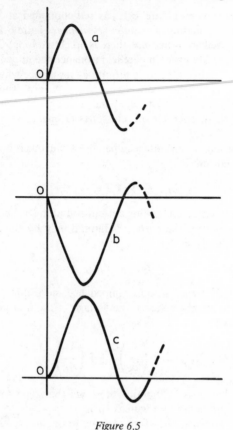

Figure 6.5

If there is some pre-loading the locus circle of I will not pass through O.

6.5 Series and parallel resonance

Resonance may be defined as the condition which arises when the effective inductive and capacitive reactances in either a branch or closed loop are algebraically equal and opposite, i.e. neglecting losses (see section 6.2 above)

$$X_L = X_C \text{ numerically}$$

and $f = 1/(2\pi\sqrt{(LC)})$ or $w = 1/\sqrt{(LC)}$ (6.6)

Parallel resonance (Figure 6.1a) is self-contained and f is the frequency of free oscillation. For *series resonance*, Figure 6.1b, there is no inherent free oscillation because there is no closed loop. Corresponding to this essential difference, in parallel resonance the impedance between a and b to an applied ac emf is infinite at resonance whereas for series resonance the impedance between a' and b' is zero. Therefore, in the absence of any circuit resistance, parallel resonance viewed externally is equivalent to an open circuit and series resonance is equivalent to a short circuit.

Further the maximum voltage in parallel resonance is line (or applied) volts and the current is

$$I_C = V/X_C = V/X_L$$

so that I_C represents a circulating current round the closed loop.

In series resonance the current is limited only by the impedance of the source of supply so that

$$I_S = -V/Z_S$$

where Z_S is the *external* or source impedance and if this is low, which is usual on power supply systems, the voltage at the mid point p (Figure 6.1b), which is

$$V_P = V\left(\frac{X_L}{Z_S}\right) = -V\left(\frac{X_C}{Z_S}\right) \tag{6.7}$$

may be very high. As in practice there will be some small losses the condition is represented by Figure 6.1(c).

On power systems, therefore, serious resonance tends to produce both high currents and high *internal* voltages, which may cause flash over, whereas parallel resonance usually produces only high currents.

When there is magnetic saturation another striking difference arises in the respective current and voltage wave forms. This is dealt with later under non-linear systems (section 6.19).

6.6 Series resonance and Boucherot circuits[8]

In a circuit such as that in Figure 6.1d series resonance occurs when $2\pi fL = 1/2\pi fC$ i.e. where $f = 1/\sqrt{(LC)}$. If in such a circuit a load, impedance Z is connected across C (or L), Figure 6.1d, then by Thevenin's theorem the current through Z is given by

$$V = (I_C + I_Z)X_L + I_C X_C = I_Z X_L + I_C(X_L + X_C)$$

but since by definition $X_L = -X_C$ (phase opposition)

$$I = V/X_L = V/X_C \tag{6.8}$$

and is independent of the value of Z.

In practice there will be some resistance, at least in the inductive arm (Figure 6.1d) and the conditions will be represented by the vector diagram in Figure 6.1c. It is clear that when the resistance R is very small in comparison with the inductive and capacitive reactances the current and voltages will be very large and would lead to flash over unless magnetic saturation of the reactance L detunes the circuit. Bartlett[8] in his paper shows that this result applies to a large class of possible circuits where at some point inductive and capacitive reactance are equal and opposite at the power supply frequency, and this might also apply to an harmonic.

6.7 Comparison with radio circuits

Consider now the conditions represented by Figure 6.6 which could cause considerable trouble. It was found that it was possible for L to

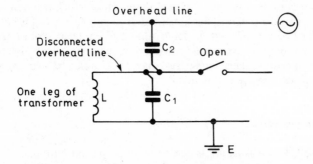

Figure 6.6 Standard oscillator circuit

resonate with C_2 which caused this winding of a disconnected transformer to become seriously overloaded. If L had resonated with C_2 there would have been serious resonance and there might have been either over current or flash over, or both.

Radio hams will, however, recognise Figure 6.6 as a standard oscillator circuit, of which many variations are to be found, and power engineers should bear this in mind when investigating unexplained

voltage troubles. Circuits of this type are deliberately used in high-frequency induction furnaces to heat-treat small metal parts and even pre-heat small steel billets for forging. R.F. oscillators for pre-heating plastics before moulding are usuallt driven from large triode power

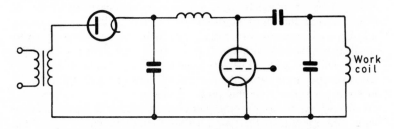

Figure 6.7 Typical power oscillator used for hf induction heating

values, but billet heaters may be driven by a tuned mercury arc. Figure 6.7 shows a typical oscillator used for high-frequency induction heating.

6.8 Current chopping

When a current is suddenly broken there will be a high reaction voltage rise $\overline{V} = -L d\overline{I}/d_t$ across the break unless as in Figure 6.1a the energy is absorbed by a condenser or, some other device such as a Silit resistance of zener diode, which will provide a momentary short circuit and relieve the stress. This voltage rise may cause arcing or sparking across the gap and prevent a clean break.

6.9 Unstable arcs

The voltage across an arc is composed of a constant term plus a second component which decreases with increasing current. One suggested imperical formula is

$$\overline{V}_{arc} = k_1 + k_2 l/\overline{I}^{\,2}$$

where l is the length and k_1 and k_2 are constants and k_1 is about 25 V. For an arc to be maintained V_{arc} cannot be less than $L(dI/dt)$. It will be seen that V_{arc}, which is of the nature of a back EMF, increasing as the gap widens, helps a circuit breaker to operate, but it increases very rapidly as I approaches zero and if L is not large the arc becomes

unstable and tends to 'snap out' and produce a voltage spike which may be very high. This is known as current chopping.

6.10 Transient voltages in transformers and reactors

The circuit in Figure 6.8 represents an idealised transformer winding. No real transformer is as simple as this and it can only lead to a qualitative picture of what occurs, but it is a very useful picture.

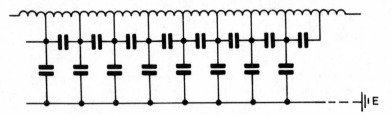

Figure 6.8 Self-capacities of ideal transformer winding

The following is based on *Surge Phenomena*, published by the ERA[3]. In the equations.

$$\phi_M = \text{main flux linking every turn}$$

$$\phi_L = \text{leakage flux linking particular turns}$$

(These are not the usual definitions used for double wound-transformers but are convenient here). ϕ_m is thus constant along the winding but ϕ_L varies. The total linkage with any turn is $\phi = \phi_M + \phi_L$

$$\text{Voltage gradient} = K \frac{d}{dt} (\phi_M + \phi_L) \text{ where } K \text{ is a function of } x$$

It follows from the definition that $\frac{d}{dx} \phi_M = 0$

for a uniform winding, therefore voltage gradient

$$\frac{dv}{dx} = K_1 \frac{d}{dt} \phi_L + K_2 \frac{d}{dt} \phi_M \qquad (6.9)$$

where K_1 is variable and K_2 is constant.

The ERA in *Surge Phenomena* is concerned with ϕ_L, but a complete solution requires both ϕ_L and ϕ_M.

The *second* term in ϕ_L arises when a large voltage such as that caused by a travelling voltage surge is suddenly applied to transformer

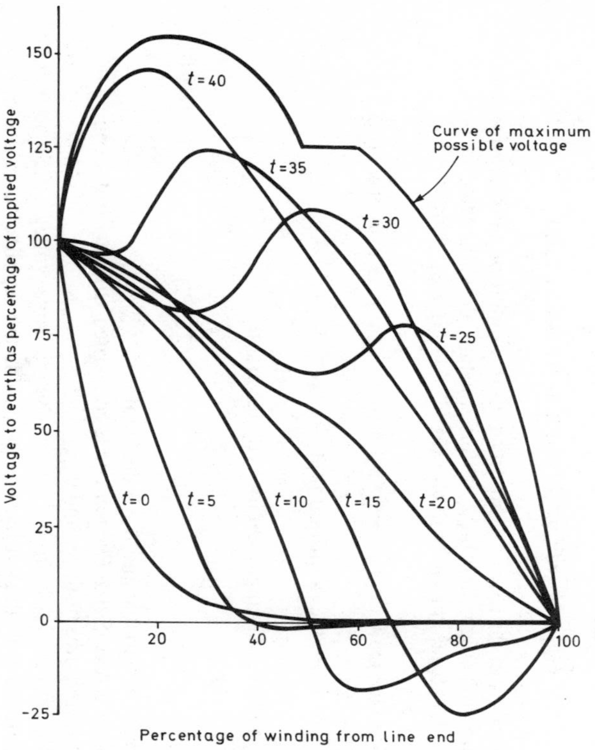

88

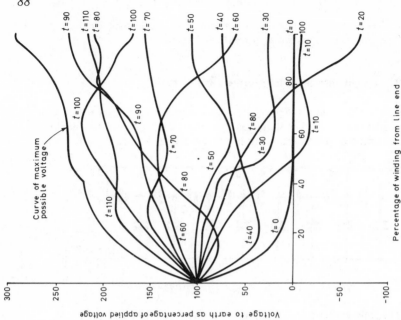

Curve of maximum possible voltage

Voltage to earth as percentage of applied voltage

Percentage of winding from line end

Figure 6.9b Voltage distribution after impact of rectangular wave (neutral point isolated). Times shown against curves are microseconds after instant of impact. (From ERA 'Surge Phenomena')

Curve of maximum possible voltage

Voltage to earth as percentage of applied voltage

Percentage of winding from line end

Figure 6.9a Voltage distribution after impact of rectangular wave (neutral point earthed). Times shown against curves are microseconds after instant of impact. (From ERA 'Surge Phenomena')

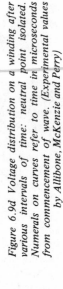

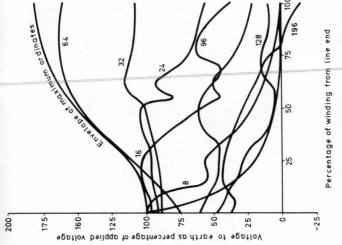

Figure 6.9d Voltage distribution on a winding after various intervals of time: neutral point isolated. Numerals on curves refer to time in microseconds from commencement of wave. (Experimental values by Allibone, McKenzie and Perry)

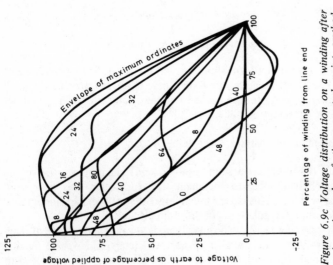

Figure 6.9c Voltage distribution on a winding after various intervals of time: neutral point earthed. Numerals on curves refer to time in microseconds from commencement of wave. (Experimental values by Allibone, McKenzie and Perry)

terminals. In the UK this has been studied experimentally by Allibone, McKenzie and Perry[9] and mathematically in the ERA publication *Surge Phenomena*. When the voltage is applied to the terminals it is immediately 'distributed' over the winding by the network of series and shunt capacitance before any current flows in the coils. The result is not, however, stable and a damped bridging transient is established which gradually subsides to the final voltage distribution. This equalising transient depends on ϕ_L only. A comparison of calculated and experimental results is shown in Figures 6.9(a) to (d). (a) and (b) show calculated results and (c) and (d) experimental ones. Although differing these are similar, the important aspect being the envelope of max voltages (They do not correspond to identical transformer designs). The voltages at points away from the supply terminal are higher than those adjacent to it. Other examples are given in Blume's *Transformer Engineering*[4] and the *J & P Transformer Book*[6]. It should be mentioned, however, that the highest voltages appear near the far end of an unearthed winding (Figure 6.9). These over voltages are normally catered for in the design of transformers by increasing the insulation locally and varying the disposition of the coil sections.

Apart from the effect of direct lightning strikes, however, the highest transient voltages probably occur as a result of current chopping (see sections 6.8 and 6.9) caused by unstable arcs.

When the current in a winding is suddenly interrupted the voltage gradient along the winding is given by

$$\frac{dv}{dx} = K_2 \frac{d\phi_M}{dt} \; app$$

(6.10)

ϕ_M being much larger than ϕ_L

thus if K_2 is fairly uniform along the winding as in the idealised case of Figure 6.8, it will oscillate like a rigid bar. This was tested with a small transformer with uniformly spaced tappings, somewhat in the manner of Allibone's tests but less elaborately[17].

For an ideal transformer, $2\pi f = 1/\sqrt{(L_M C_\epsilon)}$

where C_ϵ is the effective shunt (end to end) capacity and L_M is the inductance attributable to the main flux for which $d\phi_M/dx$ is zero along the winding. This is very much larger than the leakage inductance which determines regulation and short circuit levels.

In the case of a real transformer, there is always some degree of magnetic saturation in the interest of cost efficiency and we are dealing with a case of ferro resonance and normal a.c. theory does not lead to

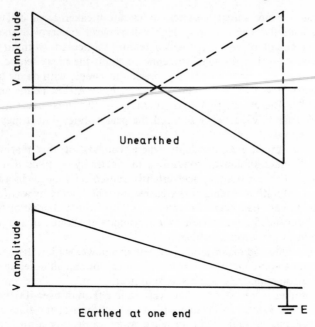

Figure 6.10 Voltage oscillations in transformer winding caused by current chopping

accurate deductions. However, the magnetising current is unlikely to be over 10% of the load current and accepting this figure

$$I_m = \frac{1}{10} \cdot \frac{kVA}{kV} \text{ r.m.s.} \qquad (6.11)$$

For simplicity consider a 1000 kVA 10 kV single phase transformer

$$I_m = 10 \text{ amp} = V/X \text{ i.e. } X = 1000 \text{ ohms}$$

$$C = \text{app } 10 \, \mu\mu F \text{ and } X_L = 2\pi f L, f = 50 \text{ hz}$$

so that $L = 3$ Henrys app

$$V_{max} = I \sqrt{\frac{L}{C}} = 0.55 \times 10^6 \text{ volts.}$$

These values are hypothetical, but of about the right order of magnitude, so that it is clear that if such a current were instantaneously interrupted a very high voltage would appear across the switch terminals. In practice this is greatly reduced because the inductive e.m.f.

prevents the arc across a switch or circuit breaker from collapsing instantaneously. Even so, very high voltages can appear with modern efficient circuit breakers, for which reason, particularly with air-blast breakers, there is a growing tendency to shunt the break temporarily with a high resistance and finally break the circuit with an air break isolating switch. (There is an interesting paper by A.F.B. Young on this subject[10]). The condition I have described is 'the worst case'; usually parallel loads or capacitance reduced the peak voltage, but it may still be very serious.

I have investigated accidents where voltages of this magnitude occurred under flash-over conditions in electricity stations. On one occasion when a workman inadvertently caused a flash-over in a large outdoor substation leading to extensive unstable arcing, it was found later that there had been a momentary flash from a conductor to an earthed expanded metal screen over a distance of about 2 m at a point 3–5 m from the original trouble.

At the time of a major fault elsewhere in a power station there was a flash from a point about one-third of the way along an air cored reactor to the surrounding screens, a distance of nearly 1 m.

At another power station there was again a flash of over 400 mm to earth from a reactor associated with 11 kV busbars at the time of an accidental busbar fault some distance away caused by a mistake in operation.

In such circumstances on a branching 3-phase supply it is rarely possible to say exactly what happened after the event – too much damage has been done, but I can suggest no other cause than current chopping at an unstable arc. It is difficult to explain why the flash-over should have occurred at a point away from the reactor terminals unless the terminal conditions required the combination of the two types of oscillation described to obtain a complete solution.

It has been suggested that these examples were the result of ionised gas from the main faults. This is improbable but cannot be entirely ruled out in the first instance; in the other cases it was impossible. In one case the reactor was in a different 'room' in a building, and in the other it was separated by about 100 m and several high walls.

Travelling voltage surges may be attenuated by surge absorbers which are designed to dissipate energy by eddy currents and/or hysteresis loss or by a sufficient length of cable. Over-voltages can be relieved by spark gaps or by various devices which are insulators at normal voltages but break down at a prescribed over-voltage. They may or may not be 'self healing'. It is good practice on overhead systems to have spark gaps or weak links one (or two) spans away from switching or transformer stations, and generally to arrange insulation levels and clearances in such a manner that if a flash over occurs it will be at a

point where minimum damage will be done. This corresponds to the use of gas and zener diodes on telecommunication and control systems.

6.11 Insulated three phase systems

The usual practice in the UK is to operate transmission and distribution systems with the neutral directly (or solidly) earthed. This effectively prevents many of the problems discussed below. However, a system may lose its earth connection under fault conditions or as a result of ill-considered switching procedures.

6.12 Arcing grounds or earths

A three-phase unearthed system under stable conditions is maintained with its neutral approximately at earth potential by balanced electrostatic capacity between lines and to earth (Figure 6.2). If, however, there is a fault to earth on one phase the other two are raised to approx. line voltage to earth. This does not necessarily impose an excessive stress on the insulation to earth of the two phases if it is of short duration. A system intended to work unearthed is normally designed with adequate increased insulation. Such a fault may, however, lead to arcing earths (or grounds) which can be very serious.

The immediate result of such a fault is that a current flows to earth at a low power factor and if there is poor contact this may cause an unstable arc to be formed, the peak voltage across which will be:

$$V_{peak} = \text{amplitude of a.c. voltage} + \text{h.f. transient}$$

If the transient oscillation is small or rapidly damped, the current may be interrupted at the power frequency current zero, leaving a trapped charge in the system and restrike with the opposite potential, thus increasing the neutral displacement and so on. If the current is interrupted during the high-frequency domain the position is similar but more complicated, and a further complication arises when the system is earthed through a reactive impedance as is the practice in the USA. The analysis of these conditions is simple in principle but complicated. Bewley summarises the results as follows, and his book may be consulted for details[1]. Clearly this can be serious.

Table 6.1
MAXIMUM VOLTAGES OF ARCING GROUNDS

Systems	Single phase	Three phase
Initial Arc: isolated neutral	3 V	2.5 V
Normal frequency arc extinction	4 V	3.5 V
High frequency arc extinction		
Isolated neutral: no damping	6 V	7.5 V
damping		5.3 to 7.5 V
Resistance in neutral		2.5 V
Reactance in neutral		3.7 to 4.0 V
Petersen coil in neutral		1.34V

V = Peaks line to neutral voltage under normal stable conditions

The voltages are not as high as those which may be caused by current chopping or lightning but they may persist for a considerable time if the capacity current to earth is not sufficient to operate selective protective gear, which is quite possible. This subject is, however, controversial and readers are referred to the paper by Wilheim and Waters 'Neutral Grounding in HV transmission' (Elsevier) which covers arcing grounds, switching surges, neutral inversion and arc suppression on Petersen coils.

6.13 'System' or neutral inversion[11]

If three conductors are connected to a transformer with an unearthed star point as in Figure 6.11(a) with *equivalent* star admittances Y_1, Y_2 and Y_3 (taken as the reciprocals of the sum of the supply and load impedances, including line capacitance)

$$\left. \begin{aligned} \dot{I}_1 &= (\dot{V}_1 - \dot{V}_p)Y_1 \\ \dot{I}_2 &= (\dot{V}_2 - \dot{V}_p)Y_2 \\ \dot{I}_3 &= (\dot{V}_3 - \dot{V}_p)Y_3 \\ \Sigma I &= \Sigma(\dot{V}_m - \dot{V}_p)Y_m \end{aligned} \right\} \text{ vectorially}$$

$$\dot{V}_p = \frac{\dot{V}_m \dot{Y}_m}{\Sigma \dot{Y}_m} \text{ if } \Sigma \dot{I} = 0 \tag{6.12}$$

The internal impedance between S and P is $1/\Sigma Y_m$ if $Z_m = 1/Y_m$ includes the transformer impedance. If, then, S and P are connected by

an external impedance $\dot{Z}_{sp}$ the current through it, by Thevenin's theorem, taking S as datum, will be

If $Z_{sp} = 0$

$$\dot{I}_s = \dot{V}_{sp}/(\dot{Z}_{sp} + 1/\Sigma\dot{Y}_m) \quad \text{(a)}$$
$$\dot{I}_s = \Sigma Y_m \dot{V}_{sp} \quad \text{(b)}$$

If S is connected to earth this represents solid earthing, otherwise the condition is equivalent to impedance (or resistance) earthing (of the neutral).

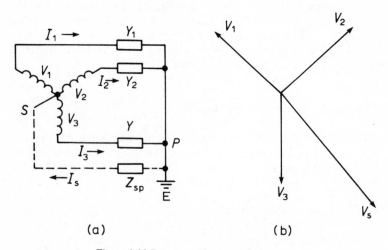

(a)   (b)

Figure 6.11 Demonstrating neutral inversion

With a normal transformer $\Sigma\dot{V} = 0$ by design and for a balanced load $\dot{Y}_1 = \dot{Y}_2 = \dot{Y}_3$ in which case $\dot{V}_{sp} = 0$. This also applies when the sole load is the phase capacitances to earth, and by the principle of superposition the conditions can be combined. In some circumstances, however, $\dot{V}_{sp}$ is so large that its vector extends outside the 'voltage triangle' (Figure 6.11b). This is called system or neutral inversion.

The virtual neutral point of a system, on the voltage phase diagram, can be defined as that point from which the sum of the phase voltage vectors is zero. With a good transformer under normal conditions this should be the star point, but under abnormal conditions this may not be true. (The formulae given in section 6.18 may be used in dealing with problems of this type.)

A very important example of inversion arises when the admittances to earth are capacitive only and one line is grounded as for an arcing ground. If the neutral has been earthed via a reactor, for example a

Petersen coil, there is a possibility of series resonance as described above and the associated voltages may be very large and dangerous. Gates[11] examined various possibilities and concluded that trouble may be experienced with

(a) Neutral-earthing transformers with one or more phases open;
(b) Petersen coils with one or more lines broken;
(c) Delta/star transformers with secondary neutral earthed and an open circuit in the delta;
(d) The condition may be a characteristic of the circuit arrangements employed, e.g. transformers without delta connected windings; star or interconnected star windings without a 4-wire supply; and with a star winding with earthed neutral connected to a cable system of appreciable capacitance.

The sudden appearance of neutral inversion may be caused by ferroresonance and the jump effect (see below).

6.14 Petersen or arc suppression coils

Because some single phase faults to earth on overhead lines such as flashover during a thunderstorm or momentary bridging of insulators by wet straws in the wind may unnecessarily cause supply to be cut off automatically and remain off until someone can visit an unattended substation, they are sometimes operated unearthed. To prevent arcing earths, described above, they have been earthed through a high impedance designed to pass a lagging current equal and opposite to the leading capacitive current from the two sound lines, so that little or no current flows through the earth fault.

As Goodlet[19] demonstrated this may cause a measure of neutral inversion. In order to prevent appreciable neutral displacement under normal operating conditions caused by unbalanced capacity currents, it is usual to detune the coil somewhat from resonance.

If, however, one phase is open circuited on the supply side (as by a broken conductor or blown high-tension fuse) the vector sum of the two capacity currents from the sound phases passes through the coil, and it works out that the neutral is displaced by an amount equal to the phase voltage (see Sloane, who recommends that the coil should therefore be tuned to a point below rather than above the resonant value of the inductance.)

Consider, however, the condition represented in Figure 6.12a which is equivalent to Figure 6.12b. It is seen that the circuit between points a and b presents the possibility of series resonance in which case anyone picking up the broken conductor at *d*, which might look innocuous, would have forced through their body, if there were perfect

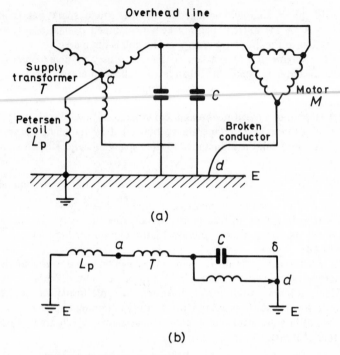

Figure 6.12 Circuit using Petersen or arc suppression coil

resonance a current approximately determined by the impedance of L (Figure 6.12b). This is perhaps unlikely, but because of the phase opposition of the currents through L and C the person would probably be subjected to a very high voltage. There are other possibilities.

Note that the shape of the BH curve has a considerable effect on the matters discussed above, e.g. series resonance, neutral inversion, arcing earths and Petersen coils; this is discussed later under the heading of 'Ferro Resonance'.

6.15 Voltage spikes in operating coil circuits

Following reports of insulation failure in low voltage contactor and relay coil circuits Russell Taylor and Randall[12] found that voltages up to 6000 to 10 000 V were produced when the inductive coil circuits were interrupted, and that when a condenser was placed across the contacts the voltage spikes were sometimes increased. The explanation appears to be as follows.

When the circuit is broken (Figure 6.12a) the current commences to charge the self and stray capacitance C, so that $d\bar{V}/dt = -\bar{I}/C$. When this

exceeds the breakdown voltage across the gap a spark passes and discharges C, the current being a heavily damped oscillation. This is repeated until the inductive energy of the coil is dissipated.

If C is augmented the passage of the spark may be delayed because $d\overline{V}/dt$ is reduced, and the breakdown voltage increases as the gap increases. The spark will then pass at a higher voltage which is limited in two ways:

    (i)   it cannot exceed the breakdown voltage

    (ii) it cannot exceed the value produced if *all* the energy is transferred from the inductance to the capacitance, i.e. since

$$\frac{1}{2}\overline{I}_m^2\, L = \frac{1}{2}\overline{V}_m^2\, C \text{ so that } \overline{V}_m = \overline{I}_m\sqrt{\frac{L}{C}} \text{ where } \overline{V}_m \text{ and } \overline{I}_m \text{ are maxima}$$

Sparking stops when (i) exceeds (ii).

We thus find that up to a point, by delaying the sparks, the voltage spikes are increased when C increases, although on further increase $\overline{V}_m$ is reduced.

In this the action of the condenser may be compared to the condenser across the contacts of a magneto or motor car induction coil.

This result has important implications for cable insulation, particularly at triforcating boxes, and for control equipment, telecommunications, solid state electronics, and intrinsic safety which are discussed in other chapters.

6.16 Instantaneous rate of rise of recovery voltage *(IRRRV)*

When a circuit breaker opens, the inductance in the circuit usually prevents the arc being extinguished until approximately current zero, which on highly reactive circuits is near the open circuit crest voltage. There is in this case a Bridging Voltage Transient the initial amplitude of which is equal to the open circuit voltage at this moment, and may be equal to the amplitude of the supply voltage; the frequency is calculated in the same way as for current transients.

Whether or not the break occurs depends on a race between the rate of rise of the recovery voltage transient and the increasing breakdown strength of the gap between the contacts – caused by cooling and deionisation. This may in simple cases be of the form:

    Recovery voltage $V = V_0\,[(A + B \sin(wt + a)]$ *

    $IRRRV = dV/dt = V_0 B\omega \cos(\omega t + a)$

    Maximum possible $IRRRV = V_0 B\omega$ or $2\pi f V_0 B$

*This represents the sum of a transient h.f. oscillation and a component which may be constant or have a low frequency compared with that of the transient.

and is therefore proportional to both the frequency and voltage. In practice conditions are often more complicated and there may be several frequencies superimposed, which depends, to some extent on factors which vary with the manner in which the system is connected at the time, so that except perhaps in major sub-stations, likely values are to a large extent a matter of experience.

A useful discussion of this matter is given by Harle and Wild[13]. For a short circuit limited by a transformer they give, for example

$$\text{mean } RRRV = 4.2fE \ V/\mu s \ (Note \ 2\pi/4.2 = 1.5)$$

where E is the 50 Hz peak voltage in kV and f is in kHz, calculated from the leakage inductance per phase and the effective lumped capacitance of the transformer windings and connecting cables. It will be noted that the maximum $RRRV$ is proportional to the product of the transient frequency and the supply voltage. [$RRRV$ is defined in British and international standards as the average, not the instantaneous value of dV/dt but $RRRV$ is difficult to calculate analytically.]

6.17 Use of stabilised arc and sparks in investigation

It is possible to stabilise a long thin ac spark (or arc) by using a voltage from a high impedance source; if the current increases the voltage drops and vice versa. This is employed in stabilising arc lamps for photocopying and I have used it experimentally to study the initial stages of developing insulation failure where the follow through arc would have destroyed the evidence; on one occasion I used an 11 kV voltage transformer excited at a considerable over voltage from the low voltage side, but with a metal filament lamp in series with the low voltage supply. The thermal characteristics of the lamp admitted a substantial initial magnetising current rush, producing a momentary over voltage on the high-tension side to cause breakdown, which was immediately choked to a lower sustaining voltage.

It is well known that heated glass is an electrolytic conductor but with high resistivity. With this apparatus I succeeded in maintaining sustained stable sparks several inches long between a pointed metal electrode and a piece of glass resting on a hot plate well below red heat and also to the glaze on a chip of electrical porcelain.

A number of flash-overs in high voltage busbar and circuit breaker enclosures appear to be associated with overheated joints in conductors. Redmayne[14] found that it was possible to establish a stable spark several inches long between a cold electrode and an artificially heated electrode (below red heat) which acted as a source of thermal electrons,

there was no high current arcing because of the rectification when the voltage reversed, but sufficient ionisation remained for re-igniting on the next reversal. In an enclosed space this would have led to flash over. We jointly studied the surface breakdown of contaminated porcelain current breaker insulator spouts in this way.

Using the same technique at 250 V I was able to watch a glowing track slowly develop across a dirty bakelised insulator, which broke down with a crack like a rifle shot when an unstabilised voltage was applied.

6.18 The absolute sum of a number of vectors

It is convenient here to introduce a simple result. If S = vector sum and A_n, A_m, etc typical vectors then

$$|S|^2 = \Sigma|\dot{A}_n|^2 + 2 \sum_{n \neq m} |\dot{A}_n||\dot{A}_m| \cos \dot{A}_n\dot{A}_m \qquad (6.13)$$

for all n and m, where A_nA_m is the angle between the two vectors. (Note: In using the result care must be taken with the signs. It is convenient to consider the vectors as originating at a single point, *all pointing outwards*, and remembering that the cosines of obtuse angles are negative.)

[Equation 6.13 is simple derived as follows:
if $S^2 = (A_1 + A_2 + A_3$ etc$)^2$ for pure numbers, by multiplying out we obtain

$$S^2 = A_1^2 + A_1A_2 + A_1A_3 \ldots \ldots$$
$$+ A_1A_2 + A_2^2 + A_2A_3 \ldots \ldots$$
$$+ A_1A_3 + A_2A_3 + A_3^2 \ldots \ldots$$
$$+ A_1A_4 \ldots \ldots$$

If $\dot{A}_1$, $\dot{A}_2$, etc are vectors, then for $\dot{A}_n\dot{A}_m$ we write the scalar product

$$|\dot{A}_n| |\dot{A}_m| \cos \dot{A}_n\dot{A}_m$$

Alternatively, but at somewhat greater length, the individual vectors may be resolved into orthogonal components and the same result follows from elementary trigonometry.]

This has a number of applications, those with which we are at present concerned being for three-phase networks. Thus, if the currents on the three phases are I_a, I_b and I_c, spaced at $120°$ (as for a purely resistive load) the out of balance current is given by

$$I_n^2 = (I_a^2 + I_b^2 + I_c^2) - (I_aI_b + I_bI_c + I_cI_d) \qquad (6.14)$$
$$= 0 \text{ if } I_a = I_b = I_c$$

since cos $I_a I_b = -\frac{1}{2}$, and similarly for the other pairs. If I_n is multiplied by the neutral impedance Z_n we obtained the displacement of the neutral caused by the unbalanced load.

6.19 Non-linear mechanical and electrical resonance

When investigating a substation bus-bar failure an examination of the debris convinced me that the short circuit had followed the fracture of supporting porcelain insulators, caused by movement of the conductors and not the other way round as had been believed. This clearly implied resonant vibrations. To confirm this I calculated the natural frequency of busbar vibrations and this was fairly accurately 100 Hz which is the frequency of the mechanical forces which would occur. About the same time another failure occurred in which the evidence was conclusive. A circular conductor about 5/8 in. dia. was severed, one half of each broken end was burned by arcing, but the other half showed a clear fatigue fracture, and again the approximate resonant frequency was close to 100 Hz. With the help of R H Sutcliffe[16], I made a series of experiments on the vibrations of busbars using the apparatus illustrated in Figure 6.13. A vibrating bar has a series of free periods, which are not harmonics because they are solutions of a fourth order differential equation; these correspond to the different modes of vibration, depending on the number of nodes and anti-nodes. But I was surprised at the closeness of the tuning; a very small change in excitation frequency suppressed resonance, and also the resonant frequency varied slightly with the amplitude. This is a characteristic of a non-linear system, the classical example being a pendulum swinging through a wide arc, for which the period is given by an elliptic integral and thus increases with increasing amplitude, i.e. the frequency decreases with increasing amplitude. If there is some damping so that the amplitude decreases, the oscillation speeds up.

Another observation was that resonance on both harmonics and sub-harmonics of the 'free' frequencies also appeared, again a characteristic of some non-linear oscillations.

It may be noted in passing that were it not for this sensitivity and amplitude dependence it would probably be impossible to repeatedly run up an alternator through a whirling speed without fatigue fracture of the shaft. As it is, there is a predetermined amplitude for any speed, which cannot be exceeded.

A non-linear system such as a pendulum may not resonate with an oscillating driving force at small amplitudes because the excitation and free frequencies do not match. If, however, it receives an impulse causing a displacement greater than that corresponding to resonance, as

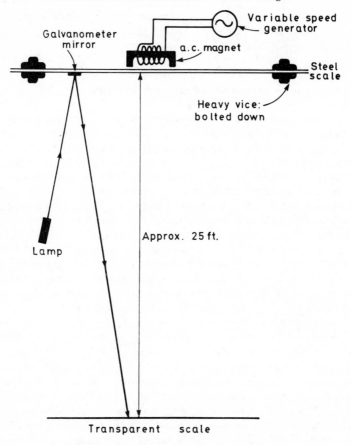

Figure 6.13 Apparatus for investigating forced vibrations. (This proved to be very sensitive)

it oscillates with decreasing amplitude it will reach the resonant frequency if the free frequency converges towards the driving frequency.

6.20 Ferro-resonance

The type of non linear resonance with which electrical power engineers are most familiar is ferro-resonance. This may arise when a larger than normal current flows through an iron cored reactance, such as a choke or transformer winding, if this causes magnetic saturation. The result is that the effective value of the inductance is reduced and therefore

$w = 1\sqrt{(LC)}$ is reduced and the coil, in association with its own or external capacitance, resonates at a significantly lower frequency than normal and the current waveform is distorted.

This is illustrated by Figure 6.14 based on work by Goodlet; the lower curve represents normal conditions and the upper curve ferro-resonance. It will be seen that if the current increases beyond A there

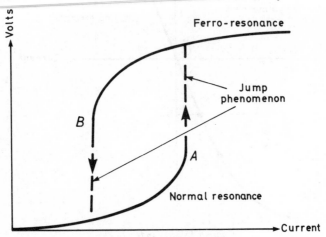

Figure 6.14 Ferro-resonance

will be a jump to the higher curve and if it is reduced below B ferro-resonance must cease. Between A and B there is an indeterminate band and conditions may suddenly change from one to the other due to external causes such as a momentary voltage spike or fluctuation, which can be very confusing. This is called the jump phenomenon.

The jump phenomenon has caused serious trouble where a system has been inadvertently left earthed via the winding of a potential transformer and it is relevant to such phenomena as neutral inversion, arcing earths and the operation of Petersen coils (discussed above). A very interesting example once occurred where the winding of a disconnected transformer was burned out. This was dealt with in detail by Wale in 1973[15].

The present tendency to use higher flux densities in transformers causes ferro-resonance to become a problem where it would not previously have occurred.

6.21 Stray magnetic fields

An aspect of ferro-resonance not covered above is that a superimposed magnetic field can cause a displacement of the range of magnetisation

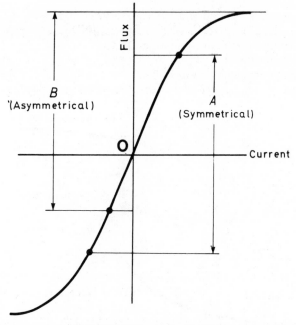

Figure 6.15 Stray magnetic fields

and thereby cause an unexpected resonance as illustrated in Figure 6.15. The mean slope of the magnetisation characteristic over range B is different to that over range A. This is equivalent to a change in the effective value of L and therefore in ω or $2\pi f$.

6.22 Potential transformers

Because of their high impedance considerable difficulty may arise in protecting potential transformers against internal faults, leading to fire or explosion. Fuse protection is not always satisfactory.

Serious trouble has also occurred when, during switching operations, or under fault conditions, a supply system has been briefly left earthed only through a potential transformer, leading to system inversion and conditions favourable to arcing earths. This has lead to serious damage and widespread shutdown, and the possibility must be considered in system design and drafting switching instructions.

6.23 Current transformers

Because of magnetic saturation, an open-circuited current transformer has a flat topped flux wave with steep sides and as a result there are

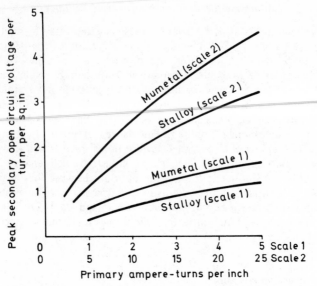

Figure 6.16 Open-circuit voltages possible with Mumetal and Stalloy.
(From Jenkins, Electrical Times 26/1/56)

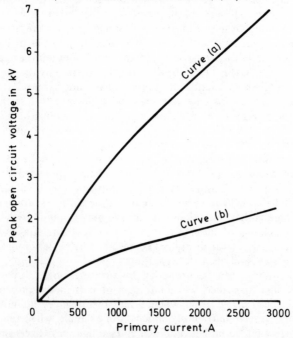

Figure 6.17 Curves showing open-circuit voltage of (a) 1000/5
ratio and 3 in² of core and (b) 400/5 ratio and 3 in² of core.
(From Jenkins, Electrical Times 26/1/56)

high voltage spikes at each current zero. The extent may be seen from Figures 6.16 and 6.17. Under short circuit conditions the peaks may easily reach 10 000 V or more, and even on full load values of up to 3000 V are quite probable. This voltage may damage insulation, and hysteresis and eddy currents may also cause damage by overheating. The greatest risk, however, probably arises from shock and, or flashover, which can occur at some distance away from the transformer if the burden becomes open circuited. The danger of open circuiting the secondary winding of an energised current transformer is well-known to power engineers, but not perhaps to industrial maintenance staff.

Current transformer rating is also important because they may not be able to withstand through faults on the primary side. This can now be dealt with by stating withstand values for I^2 and $[I^2 t]$. The windings have in some cases burst on through faults and mechanical integrity is important (see section 7.5).

6.24 Prevention and cure

In this chapter we have dealt with a number of causes of abnormal voltages. Some could have been prevented in the design stage.

For example, system inversion and arcing grounds are avoided by solid earthing of the neutral, which is usually possible in industrial installations supplied from a works transformer, but with several transformers or generators in parallel it is usual to earth only one neutral to prevent circulating currents, and in particular third harmonics, flowing between them. In this case care must be taken not to lose the earth connection as a result of switching procedures, including operation on faults. It is particularly dangerous to leave a system earthed only through a potential transformer primary winding.

Many voltage spikes caused by current chopping may be avoided by a proper choice of cartridge fuses and others suppressed by shunt connected silicon or other semi-conductors for power circuits, and zener diodes for light current circuits. Unless this is done mineral insulated metal sheathed cables in particular may be punctured.

The possibility of flash over in high voltage trifurcating boxes on transformer, switchgear or motors is prevented by better insulation, filling techniques, and/or phase separation.

Some mishaps can be prevented by never leaving a disconnected overhead line unearthed when any part of it is in the proximity of other overhead lines. Even when severe disturbances do not occur people may receive fairly nasty shocks. For example, when a bus (with rubber tyres) pulled up below a high voltage line, passengers getting in and out received very unpleasant shocks, and at a Yorkshire power

station a man working on a dead outdoor 66 kV overhead bus-bars or isolators received a violent shock when opening an isolator removed the earth connection and he sat down very hard on the steel gantry. Fortunately he did not fall off.

Fortuitous resonance effects can be suppressed either by detuning or damping. Generally speaking, when obscure over-voltages arise the greatest difficulty is to decide on the cause which can be very obscure, and a proper understanding of the underlying theory is essential. Once found, the cure is usually obvious and simple, but a wrong diagnosis may prove costly.

References

1. Bewley, L.V., *Travelling waves on transmission systems,* John Wiley & Sons (1933)
2. Dannent, C. and Dalgleish, J.W., *Electric power transmission and interconnection,* Pitman (1930)
3. E.R.A. *Surge Phenomena,* Seven Years Research for CEGB (1933–40)
4. Blume, L.F. (editor), *Transformer engineering,* John Wiley & Sons (1938)
5. Minorsky, N., *Non-linear mechanics,* J.W. Edwards (1947)
6. Stigant, S. Austen and Franklin, A.C., *'The J & P transformer book',* 10th edition, Butterworths (1970)
7. Smith, Charles H., *Elementary treatise on conic sections,* MacMillan & Co. (1910)
8. Bartlett, A.C., 'Bucherots constant current circuits etc', *Jour IEE* **65** (March 1927) (Bucherot Rev. Gen de l'Elect V.5 1919)
9. Allibone, T.E., McKenzie, D.B. and Perry, F.R., 'Effects of impulse voltages on transformer windings'. *Jour IEE* (1937)
10. Young, A.F.B., 'Some researches on current chopping in high-voltage circuit breakers'. *Proc IEE,* **10,** p. 337 (1953)
11. Gates, B.G., 'Neutral inversion in Power Stations', *Jour IEEE* **78** (1936) Mortlock, J.R., 'Earthed potential transformers on insulated systems', *Elec. Times* (Jan 1944)
12. Taylor, J.R. and Randall, C.E., 'Voltage surge caused by contactor coils', *Jour IEEE* **90** Pt II (1943)
13. Harle, J.A. and Wild, R.W., 'Restriking voltage as a factor on the performance, etc. of circuit breakers', *Proc IEEE* **90** Pt II (1944)
14. Redmayne, H., *Private communication*
15. Wale, G.D., 'Ferro resonance in a disconnected e.h.v. power system', *GEC Jour of Science & Technology* **40** 2 (1973)
16. Experiments by W. Fordham Cooper and R.A Sutcliffe.
19. Goodlet, 'A note on voltage instability in testing equipment', *Jour IEE* **80** p. 490 (1937)

Chapter 7

Excess Currents and Excess Current Protection

The list of symbols given in Chapter 6 is also relevant to this chapter.

7.1 Current surges

In the previous chapter the effect of bridging transients and the possibility of very large currents flowing when there is series resonance, which is equivalent to a short circuit if the resistance is low, has been described. There is also a possibility of momentarily excessive currents when certain equipment is switched on.

Technically, the simplest current surge arises when large banks of filament lamps which have a much lower resistance cold than when hot are switched on. This can be a nuisance because large lighting loads may blow fuses or trip overload protection. It is possible technically to express this by saying that the let through $[I^2 t]$ (described below) must be related to the withstand values of the protection. This can be dealt with in the light of experience by increasing settings and fuse sizes. It does, however, militate against close protection.

A more troublesome cause of current surges arises when large banks of condensers are switched. If, for example, a circuit is energised at the moment of maximum voltage a very large charging current will flow, particularly if, for some reason, the condensers have retained a charge of the reverse polarity. Considerable trouble has been experienced with banks of condensers when switched for power factor correction, or tuning on high frequency furance installations, causing heavy equalising currents to flow between sections in the absence of appreciable resistance or reactance in the connections. To prevent damage large h.r.c. fuses are sometimes placed between sections. On electricity supply networks it is well-known that switching unloaded or lightly loaded cable networks creates special difficulties in the use of circuit breakers. This is a matter which should be discussed with the specialist

108

staffs of manufacturers or supply authority transmission engineers and is outside the scope of this book.

Switching large power transformers may also cause large current surges if closure occurs near maximum voltage and the transformer core has been left with high remanent magnetism, since this means that its effective impedance is greatly reduced in the vicinity of saturation, and

$$V = \frac{d}{dt} LI \quad not \quad L \frac{dI}{dt}$$

The latter is only true on that part of the characteristic where L is independent of I. This matter is dealt with in detail in such works as *'The J & P Transformer Book'*[1] and is well-known to transmission and protective gear engineers.

Another cause of trouble arises in starting up motors driving high inertia loads. At low speeds, induction motors in particular draw very large currents from the line. There is always a brief overload, which prevents close overload protection during starting, but with high inertia and a slow start this becomes unduly prolonged and may overload the equipment or the supply cables. It is therefore important in such cases to make certain that the motor and starter are both suited to the duty. Failure to do this may cause the motor to burn out or the starter to explode.

7.2 Estimation of potential short circuit levels in sub-stations

When purchasing high voltage circuit breakers, professional advice is usually obtained either from consultants or manufacturers, unless staff engineers have special experience of these matters. For major substations approximating to large grid substations expert advice is essential. Before embarking on a new installation, or overhauling an old one, it may be desirable, however, to make an approximate assessment of requirements, and the following notes are provided bearing this in mind.

Circuit breakers are rated, in kVA, on the system voltage, ampere carrying capacity and short circuit breaking capacity. The breaking capacity has no essential relation to the load current but depends on the power available from the supply terminals which can be pumped into a short circuit. The breaking or rupturing capacity is normally stated as the maximum symmetrical short circuit kVA potentially available, with which the circuit breaker can cope. This is not the whole story but it is not usually necessary to go any further in assessing the requirements for an industrial installation.

The short circuit level at the incoming high voltage terminals can be

obtained from the electricity authority, and this is the value ᴜ.at should be specified for circuit breakers to be installed at this point. The authority should be able to quote the supply impedance per cent for use in specifying the duty at other points.

Transformer impedances are stated on their name plates as copper loss % and Reactance % based on the transformer capacity. Cable and overhead line impedances can be reduced to percentages based either on the capacity of the transformer or an arbitrary value such as 1000 or 10 000 kVA by the following formula:

$$Z\% = \frac{\text{Base kVA}}{10(\text{kV})^2} \times Z \text{ in ohms}$$

This can be made clear by the following example (see also Figure 7.1) in which it is assumed that the supply impedance is so small compared

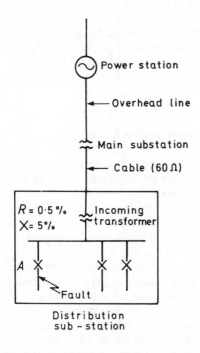

Figure 7.1 In this diagram show-ing a short-circuit, the necessary rupturing capacity of circuit breaker A is required

with that of the incoming transformer bank that it can be neglected. All % values must be reduced to a single base value, and 1000 kVA is convenient here.

Cable resistance % on basis of 60 ohms and 1000 kVA at 11 kV

$$= \frac{1000 \times 60}{10 \times 11^2} = 5\% \text{ approx}$$

(*Note*. 60 ohms is rather large but is given purely as an example to illustrate the use of the formula.)

Total impedance at switchboard:

$$Z\% = 0.5 + 5j \quad \text{transformer}$$
$$\underline{+5.0 + 0j} \quad \text{cable}$$
$$= 5.5 + 5j = 7.4 \text{ approx. i.e. } (5.5^2 + 5^2)^{\frac{1}{2}}$$

Rupturing duty

$$= \frac{1000 \text{ kVA}}{Z\%} = \frac{1000 \times 100}{7.4} = 13\,500 \text{ kVA approx.}$$

The advantage of using the percentage method is that no account need be taken of transformation ratios, the transformer being treated as a series impedance; and the arithmetic is the same for single and three-phase supplies. Percentage impedances may be combined for series and parallel circuits just as with ohmic values.

As previously stated, this is not the whole story and more detailed calculations are required for important supply substations. Also at the highest voltages, circuit breakers need virtually to be tailormade for the site and duty. A convenient account of such calculations is given in considerable detail in the *J & P Switchgear Book*[2].

Sometimes impedances are added numerically instead of vectorially as above. The error may not be great but by taking note of the phase angle a better idea of the stress on a circuit can be obtained since a low power factor current is more difficult to break.

7.3 Rupturing duty and causes of switchgear failure

A circuit breaker must be able to interrupt the maximum possible short circuit current without serious damage, unless it is backed up by another device (e.g. h.r.c. fuses) as described below, and also to carry and make or break the rated current and normal overloads over long periods without distress. As explained in Chapter 2, rupturing capacity is a statistical property and in principle a very large number of tests would be required to establish it conclusively. In practice a reasonable

number of type tests suffice to demonstrate that it is very improbable that a circuit breaker *in good condition* will fail in service.

Failures of recent designs caused solely by inadequate rupturing capacity are rare, but disastrous explosive failures caused by lack of maintenance, wrong application, or abnormal conditions may occur and these very frequently have mechanical or metallurgical causes, some of which have been described in Chapter 5.

The following are typical examples of what can go wrong with oil circuit breakers used on industrial premises, i.e. when used at up to 11 kV (or possibly 33 kV or 66 kV in very large installations). When an oil circuit breaker fails to clear a fault a more or less severe explosion usually follows. But this is not the only or perhaps the most important cause of explosions.

In heavy industry, such as steel making or rolling an 11 kV oil circuit breaker may be used in effect as part of the control gear of a large motor or process and it will therefore be operated much more frequently than would have been the case in a public supply substation. An extreme example was the control of arc furnaces in the early years of the 1939 war when maximum production was essential and routine maintenance became difficult. Circuit breakers which were designed for operation perhaps a dozen times in a year had to operate 20 or 30 times a day and fatigue and impact failure, caused by the sudden deceleration of the moving parts as the contacts closed, showed up unsuspected design weaknesses; for example, on several occasions the cross arm fell off and was found in the bottom of the tank; on other occasions sparking contacts fell off. These failures fortunately did not cause many serious accidents and were traced to 'stress raisers' (see Chapter 5), weak dowel pins and inadequate locking of studs or nuts.

A more serious fault was oil contamination caused by frequent operation. Arcing under oil causes the formation of fine or colloidal carbon and explosive gases such as hydrogen, methane and acetylene. During periods of rest carbon particles settling on insulation may cause a tenuous conducting path followed by flash over, with or without tracking. It is imperative, therefore, that the oil should be replaced at short intervals and the insulation wiped clean if serious explosions are to be avoided. At one steelworks the oil was continuously cleaned by circulating it through an automobile oil filter by means of a small pump. Nowadays this trouble has been reduced by the introduction of air break and air blast, high voltage circuit breakers.

Oil contamination may, however, still occur following the clearance of very heavy short circuits. After such an occurrence the tank should be lowered, the contacts filed to remove copper beads and other arc damage, and the oil changed. This should be done as soon as possible. During the second world war in 1940 some substation circuit breakers

were called upon to clear several heavy faults in a short period of time, which they did successfully, but under the circumstances immediate attention was impossible. One circuit breaker exploded an hour or two after its last operation by which time carbon had settled. During the period of emergency each operation stirred up the oil and prevented carbon settling out, but at the same time increased the amount in suspension to settle later.

A somewhat similar result occurred at a substation in a hilly rural area during a thunderstorm. One circuit breaker cleared several overhead line faults and each time was re-closed. After the storm it exploded and completely destroyed the brick substation.

Other explosions have occurred with older designs of switchgear by the spring of contacts weakening and reducing the contact pressure, causing a continual stream of small arcs and sparks and the evolution of inflammable gas. A similar failure at the plug contacts of a bus-bar voltage transformer caused another explosion.

The explosion may, however, occur outside the circuit breaker. In the early days of the national electrical grid a large volume oil circuit breaker (either 33 kV or 66 kV) cleared a fault successfully but an external explosion bulged the 14-inch brick substation walls and let down the reinforced concrete roof on to the top of the switchgear. This switch chamber was about 30 foot × 60 foot by 20 foot high. The circuit breaker itself was virtually undamaged and had only lost a small amount of oil. The explanation appears to be that the momentary rise of pressure inside the tank elastically stretched the tank securing bolts a little and a mixture of hot gas and oil was ejected which ignited on contact with the outside air.

Another, smaller external explosion once occurred when explosive gas collected in the space between the circuit breaker top and the bus-bar chamber on some switchgear with downward isolation. It was ignited when the oil circuit breaker was lowered for cleaning.

As examples of what can go wrong, the following may be mentioned:

From time to time spanners, nuts, etc may be dropped into switch tanks, and may not cause damage but can cause short circuits. After one explosion, however, a cleaning rag was found wedged between the switch contacts.

On another occasion an oil circuit-breaker tank had been dropped to replace the oil, but was replaced empty. The circuit breaker was safely closed but exploded the next time it was opened.

All oil circuit-breaker explosions are potentially very dangerous. The cure lies in attention to the finer points of design, particularly mechanical detailing, and of course systematic maintenance, and oil cleaning after operation on faults.

The use of air-break and air-blast breakers reduces the chance of

serious explosions. One inherent difficulty with high voltages is the necessity of providing large air break isolators, commonly in the form of long rotary knife switches. These when arrested in the open or closed position can introduce shock stresses because of their high inertia and this, as expected, caused failures with early designs, as also occurred with steam expansion breakers. I have also heard of at least one explosion of a porcelain insulator subject to internal air pressure.

7.4 The operation of fuses

This discussion is related to a.c. circuits. Conditions with d.c. are more demanding, but are relatively unimportant in industry.

Single-wire (rewirable) fuses are the oldest devices for protecting a circuit against excess current. They seem so simple that hardly anyone gives them a thought beyond ascertaining that they have the correct current rating. In fact, however, the mode of operation of fuses is extremely complicated physically and they need to be designed for specific duties. With modern electrical equipment it is essential to ensure that a correct type of fuse is employed.

The duty which a fuse may have to perform depends on a number of uncontrolled circumstances. In standards it is therefore necessary to specify its rating in a somewhat arbitrary manner to achieve reproduceable test results. In the following discussion I have not followed BS 88 (the basic specification) in detail but have tried to relate its terms to the needs of the user rather than the manufacturer or test house. As a result terms may not be defined in exactly the standard way.

7.5 Rapid heating of conductors [$I^2 t$]

When a conductor is heated by a current so rapidly that virtually no heat is lost, the rate of rise of temperature is given by

$$\frac{d\theta}{dt} = i^2 \rho/c$$

$$\text{or} \quad i^2 \, dt = \frac{c}{\rho} \, d\theta$$

where t = current density;
$\quad \rho$ = specific resistance;
$\quad \theta$ = temperature;
$\quad c$ = specific heat.
$\quad \alpha$ = temperature coefficient

ρ is not a constant but is given by $\rho = \rho_0 (1 + \alpha\theta)$ approx, hence

$$\int_0^r i^2 \, dt = \frac{c}{\rho_0} \int_{\theta_1}^{\theta_2} \frac{d\theta}{1 + \alpha\theta} = \frac{c}{\alpha\rho_0} \log h(1 + d\theta) \tag{7.1}$$

Therefore the conductor melts at a fixed value of $\int_0^r i^2 \, dt$ which is a physical property of the material and which can be measured or calculated.

In the following discussion $\int_0^r i^2 \, dt$ will be written conventionally as $[I^2 t]$, the value of which integrated up to the moment of melting of a fuse is usually called the *pre-arcing* $[I^2 t]$.

It will also be noted that $[I^2 t] R$ is the heat generated in any external conductor such as a cable or winding and $[I^2 t]$ is therefore a measure of the damage which may be done by a short circuit, when the integration is taken over the complete operation including both pre-arcing and arcing time. This is an important quantity and is called the *let through* $[I^2 t]$. Because $[I^2 t]$ may be defined as the energy dissipated as heat during a specified operation, per ohm of resistance, it is often called the *specific energy* of the operation.

7.6 Fuse nomenclature

The *size* of a fuse refers to its rated carrying capacity, not to its breaking capacity or its physical dimensions.

The operation of a fuse in an a.c. circuit is illustrated in Figure 7.2. A distinction must be drawn between the *potential current* and the *nominal current* obtained by dividing the r.m.s. volts by the impedance (but see note on prospective current below). The peak of the a.c. component is 1.41 times the normal or r.m.s. current, but the potential maximum current includes any transient d.c. component and may be nearly twice as large again, so that the maximum possible potential current, allowing for decrement, will be not quite 2.8 times the nominal current. The value of the potential current depends, however, on the *point on the voltage wave* at which the short circuit commences. (This is often contracted to 'point-on-wave'.)

The *cut-off current* is determined by the pre-arcing $[I^2 t]$ integrated up to the moment of melting. (For all ordinary purposes melting coincides with the commencement of arcing although there is a very small delay.)

The cut off is most effective when the size of a fuse is small compared with the potential current. For a relatively large fuse the cut-off current may approach the magnitude of the potential current and be

larger than the nominal current, which is quite straightforward but may at first glance seem absurd.

The *current rating of a cartridge fuse link is a value assigned by the manufacturer*, i.e. the current which the manufacturer guarantees the fuse will carry continuously under normal conditions, without either

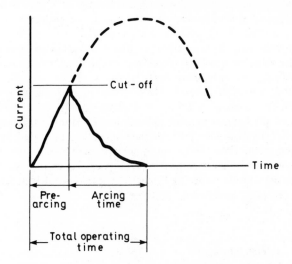

*Figure 7.2 Graph of time/current relationship for the opera-
tion of a fuse in an a.c. circuit*

blowing or deteriorating. (If the fuse is used in a hot unventilated position it may need to be derated.)

The limiting effect of the cut off on a 450 V distribution system would be much greater for a 5 or 20 A rated fuse than a 500 A, although their nominal breaking capacity may be the same (see below).

The *fusing factor* should be the ratio of the maximum current which the fuse will only just carry indefinitely long without blowing, to the current rating. For practical purposes in testing it is defined as

$$ff = \frac{\text{four hour blowing current}}{\text{current rating}} \qquad (7.2)$$

The IEE wiring regulations state in effect that for *close protection* the fusing factor must not exceed 1.5; otherwise the protection is called 'coarse'.

Prospective current is defined in BS 88 as the r.m.s. value of the a.c. component of the current, which would flow if the fuse link were replaced by a solid link of zero resistance. It is defined in this way to

minimise disagreements over the results of acceptance or proving tests. Prospective current is virtually the same as nominal current which I have defined from the point of view of the applications or plant engineer.

The breaking capacity is the maximum prospective or nominal fault current of a circuit for which a fuse should be used. The actual potential current depends on the 'point on wave' at which the short circuit commences and may be 2.8 times the prospective or nominal current.

Fuse characteristics are curves in which operating time is plotted against current broken. These are important in ensuring discrimination between fuses and between a fuse and electromagnetic or other overload protection (see below). This presents no problems for slow operation on overload but the time of operation on short circuits depends on variable factors including in particular the 'point on wave' at which the fault commences and is difficult to determine[3]. For this reason it is common practice to plot the prospective current against the virtual time defined as

$$\text{Virtual pre-arcing time} \ = \ \frac{\text{pre-arcing } [I^2 t]}{(\text{prospective current})^2} \qquad (7.3)$$

(*Note:* The pre-arcing $[I^2 t]$ is fixed by the fuse design; the total let through $[I^2 t]$ depends also on external factors, but as a rough estimate may, I suggest, be taken as approximately twice the pre-arcing $[I^2 t]$).

In the past the above refinements were important only to fuse designers and test laboratories but with increasing sophistication and in particular the use of solid state electrical devices in telecommunication, instrumentation and control systems they are becoming very important to applications and plant engineers.

7.7 Elementary principles

When a wire is slowly heated by current the clamps at the ends absorb heat, so that the centre is the hottest point and melts first. If, however, a very high current is passed through it so that it heats up very suddenly and very little heat is lost, it becomes uniformly hot except very close to the ends and virtually the whole surface reaches melting point at the same time. When there is a sufficient depth of molten metal the wire breaks up into unduloids; this is controlled by surface tension, the magnetic pinch effect and viscosity.

Neglecting viscosity, Lord Rayleigh[5] calculated in 1878 that the length of the unduloids on the surface of a cylinder of liquid should be

about 4.5 times the undisturbed diameter, which fits fairly well. When the unduloids develop into separate droplets a small arc forms between each pair so that the number of arcs in series is proportional to the length of wire and the voltage required to sustain these arcs will also be proportional to the length. This occurs at a fixed value of $[I^2 t]$ for a particular design of fuse. The whole process takes place in a few milliseconds or less.

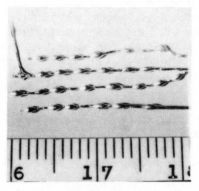

Figure 7.3   Unduloids of a blowing fuse (silver wire).
Scale in cm. (ERA)

To sustain the arc it is necessary in the case of slow heating, for the circuit voltage to exceed a value which increases as the length of the arc increases, and extinction will occur at a current zero when the voltage available is insufficient to cause it to restrike.

In the case of a short circuit a string of small gaps is suddenly produced which are bridged by arcs and the current rapidly falls. The sustaining voltage required, which is proportional to the number of arcs and therefore to the length of the wire, is provided by the circuit voltage plus the inductive voltage $L \dfrac{dI}{dt}$ produced by the fall in current. The circuit will be broken when this is less than the sum of the arc voltages and this usually occurs *within* the first current loop. There is a tendency, therefore, for the volts to rise to a value proportional to the length of the fuse wire and if damage is to be avoided a fuse must not be too long. This is particularly important where solid state electronic devices may be involved.

7.8 Arc extinction

The voltage across an a.c. arc as its length increases by vaporising the metal is indicated in Figure 7.4. It will be noted that the centre of each

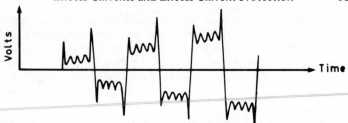

Figure 7.4 Volts across a.c. arc, increasing in length with time (note voltage spikes at start and end of each pulse)

pulse is roughly constant but that there is a spike at each end. This may be represented by, say

$$V_a = V_c + Al/(B + l)$$

where l = length;
A and B are constants.

Immediately before current zero

$$V_a = V_0 + \frac{A}{B} l$$

which increases with the length of the arc. For a long fuse under short circuit conditions the value will be $V = nV_a$ where n is the number of arcs in series. This will be balanced by

$$V = V_c + L \frac{dI}{dt}$$

where V_c is the circuit voltage at the moment of arc extinction.

Because of the presence of inductance V may substantially exceed the normal a.c. peak voltage. For this reason there is a maximum permissible length of fuse wire. This is particularly important when there are solid state circuit elements. It is a special case of current chopping.

There is some evidence, however, that circumstances occasionally arise where there is a premature collapse of the current which is much more rapid and very high voltage spikes occur. Otherwise it is difficult to explain the occasional flash-over inside well-designed ironclad fuseboxes with no apparent failure of the cartridge fuses in them.

7.9 Withstand and let through values

This concept has been increasingly used in defining the protection needed for electrical apparatus and circuits. It has been shown above

that for excess currents of very short duration $\int I^2 \, dt$ (commonly referred to as the $[I^2 t]$ value) taken over the appropriate interval determines the maximum temperature reached. This has several aspects which may be defined as follows:

1. There are fixed values for the short circuit pre-arcing $[I^2 t]$ for each design of fuse, and for the minimum, long time, operating current.

2. Integrated over the pre-arcing plus arcing time of the fuse operation it gives the *let through* $[I^2 t]$ which determines the amount of overheating suffered by the connected circuits etc.

3. Conversely, each cable and apparatus will have *withstand* values of both current and $[I^2 t]$ which must be matched by the operating characteristics of the protective equipment, whether fuses or circuit breakers.

Similarly, apparatus and cables have withstand values for over-voltages, particularly those of very short duration, against which they must be protected. These may arise from a variety of causes, such as travelling surges caused by switching operations or lightning or even the breaking of relay coil circuits. Protection is usually provided by special devices such as surge absorbers, or spark gaps and semiconductor devices such as Silit bridges and more recently zener diodes.

Withstand voltage values are important in the application of mineral insulated metal sheathed cables, (which have lower withstand values than rubber, for example) for voltage spikes, and for semiconductor devices used in telecommunications and control systems, which are very sensitive to over-voltages. Withstand values are also of great importance in relation to the barrier devices used on intrinsically safe circuits for danger areas. Unfortunately, fuses themselves may in some circumstances cause voltage spikes by current chopping.

At the other extreme, cables and apparatus are able to withstand lower currents for a very long time without serious danger. This may be called the overload or continuous withstand value, and this depends on I^2 not $[I^2 t]$. It is usually easier to use magnetic overload devices rather than fuses, because these can be set to operate at a fixed current value and a substantial margin of error is not required. However, fuses are cheaper, require less maintenance and are therefore frequently used when accurate settings are not necessary.

Between the two extremes of permissible over-currents there is an intermediate region represented by time-current characteristics, generally presented as graphs, for fuses and electromagnetic devices (Figure 7.5). These characteristics should correspond to the thermal withstand characteristics of the apparatus protected. The protection of a cable, for example, may need a different characteristic to that of a motor.

If a device will protect a circuit at both ends of the scale, i.e. as to

let through [$I^2 t$] and continuous overload, it will usually be adequate in between, although there may be exceptions. Often, however, no single form of protection will suffice and on medium voltage systems a switch or circuit breaker with an electromagnetic release is used for overloads, backed up by h.r.c. fuses to give protection against short circuits. The point of crossover of their characteristics then becomes

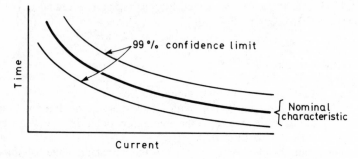

Figure 7.5 Graph of time/current characteristics for fuses or electromag-netic devices

important. This duplication often arises because a switch or controller, which can deal with overloads, would have to be excessively large and expensive if it had to cope with short circuits (see section 2.16 and Figure 2.11).

This is a controversial matter on which more research is necessary and is beyond the scope of this brief note. (See ERA Report 5012.)

7.10 Serial discrimination

On a branching system or a long feeder with loads fed off at different points along the route, the fault or short circuit values reduce as the distance increases. It is important that a circuit breaker or fuse 'near home' shall not operate when a small fault occurs at the boundary and thus disconnect a larger part of the system than is necessary. For this reason either the time/current characteristics of the protective devices must not cross, so that the smaller, remote, device at the periphery always operates first, or the characteristics must cross over in a manner strictly related to the withstand values of the items protected, so that no part is disconnected unnecessarily. However, because it takes a measurable time for any protective device to operate, it is important that the 'home' protection shall not be committed to breaking the circuit before the outer device has cleared the fault, and allowance

must be made for the statistical variation between the characteristics of different items made to the same specification (see Chapter 2).

This matter is discussed below, but it is clear that the total *let through* $[I^2 t]$ of a remote fuse must be less than the pre-arcing $[I^2 t]$ of the home fuse with a margin of error because a fuse heated very near its melting point may have its carrying capacity and future let through $[I^2 t]$ reduced. The effect is, however, small (see Turner and others, ERA Report 5012). The preheated fuse also has a reduced let through $[I^2 t]$ (see ERA Report GT-330).

7.11 Basic requirements

A fuse with its holder, terminals, etc must:
- (i) carry the load current without overheating or deterioration;
- (ii) suppress or interrupt the maximum potential short circuit current without distress;
- (iii) limit the short circuit $[I^2 t]$ to the withstand values of apparatus and cables it should protect (cut off);
- (iv) limit the sustained I^2 to the withstand values;
- (v) not cause a voltage rise in excess of the withstand values when it operates;
- (vi) in doing any of the above it must not overheat, explode, scatter flame, hot gases, metal vapour or droplets or otherwise cause danger. In the case of cartridge fuses any damage must not prevent their correct replacement;
- (vii) its time/current characteristic must match with other fuses circuit breakers, overload relays, etc so as to ensure correct discrimination, i.e. so that two forms of protection do not operate in the wrong order;
- (viii) a fuse cartridge or its holder, or the holder of a semi-enclosed wire fuse should not fit a base or holder on a circuit for which it is unsuitable;
- (ix) it must be possible to replace the fuse (or holder) without the risk of touching live metal (except on very low voltages) or causing a short circuit or, on control circuits, causing danger in any other way;
- (x) it must be suitable for the voltage of the circuit in which it is used.

Some of these requirements are considered below. Other requirements relating to particular equipment, such as cables and their ratings are discussed in the relevant chapter. The operation of a fuse on short circuit may take less than 1/500 sec and is difficult to investigate, but in recent years it has been found possible to calculate its behaviour.

7.12 Time/current characteristics

When the time taken for a fuse to operate (or 'blow') is plotted against
the current it carries a curve, such as shown in Figure 7.6. But for high
currents and short times this is insufficient and often we need to know

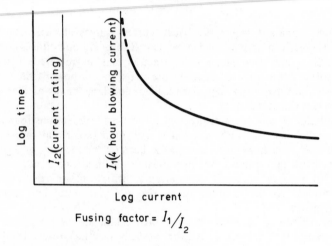

Fusing factor = I_1 / I_2

Figure 7.6 Graph showing effect of time/current characteristics

the time up to melting or arcing, and the arcing time before final extinction, Figure 7.2. Fuse operating time is, however, subject to statistical
variations and different fuses to the same specification will behave
slightly differently so that the characteristic should be represented by
confidence limits as in Figure 7.5.

The first requirement is that the fuse should carry the rated current
without overheating or deteriorating. This latter point is particularly
important for open and some enclosed fuses but is less likely to be a
limiting factor when cartridge fuses are used. The characteristic should
match the withstand characteristic of the cables or apparatus protected.
The rate of temperature rise of a transformer, motor and cable may
differ substantially and if close protection is required, different fuse
designs may be necessary. The curves representing the current time
characteristic of the fuse should lie above, but as close to that representing the thermal withstand characteristic of the apparatus as is
practicable but a margin of safety is necessary to prevent nuisance
operation. The margin allowable for a rubber cable is usually greater
than for a thermoplastic cable such as pvc and the margin for both is
much greater than for some semiconductor devices.

At present four classes of low and medium voltage fuses have general recognition:

Class P : fusing factor up to 1.25
Class Q_1 : fusing factor up to 1.5
Class Q_2 : fusing factor up to 1.75
Class R : fusing factor up to 1.75

Some apparatus may take heavy current surges at switching on. These surges can only be carried by Class R fuses although this is coarse protection, and even larger fusing factors may be required. Typical examples of this type of apparatus are transformers, power factor condensers, motors with a high inertia load, and large banks of filament lamps (see Chapter 13).

In order to produce cartridges with very small dimensions for some continental back up fuses associated with miniature circuit breakers, performance on small over-currents has been sacrificed. They give good short circuit performance but they must not be used where they would have to operate on small over-currents. British 'motor' rated fuses compromise on the need for small size and give better low current performance.

Semi-conductor devices, on the other hand, are very sensitive to over-loading and these may require specially designed fuses. There is thus a continuing conflict between the necessity to have specially designed fuses for particular purposes and the need to standardise as few types as possible for mass-production. The IEC is at present striving to produce optimum international standards.

7.13 Let through $[I^2 t]$ and back-up protection

The concept of $[I^2 t]$ was first introduced as a means of specifying the severity of fuse tests, but it is increasingly used as a criterion of the degree of protection against thermal damage afforded to protected apparatus, in particular semi-conductors. C S Lawson[3] in 1957 gave the following comparison of small fuses with miniature circuit breakers on medium voltage circuits.

Current rating (A)	Permitted amp–sec Fuses	m.c.b[s]
5	30 to 50	
10	100 to 200	
15	250 to 400	30 000 to 320 000
20	500 to 800	
30	1000 to 1500	

He further stated that

15 000 amp-sec will melt a 14/0.0076 flexible cable
40 000 amp-sec will melt a 23/0.0076 flexible cable
150 000 amp-sec will melt a 3/0.029 flexible cable
310 000 amp-sec will melt a 7/0.029 flexible cable

Cable insulation will, however, be seriously damaged by overheating long before the conductors melt, a partial exception being mineral-insulated metal-sheathed conductors.

It is now generally agreed that miniature circuit breakers (which are further considered in Chapter 8) may need to be backed up, when the prospective short circuit current is high, by h.r.c. fuses which will operate on short circuit before the circuit breaker can open. This is partly because small h.r.c. fuses cut off the current long before it reaches the potential maximum and also because m.c.b's may have insufficient breaking capacity to deal with heavy short circuits and may fail explosively. This is unlikely to happen on domestic, small office or similar systems, or at remote points on industrial systems where the short circuit current is limited by cable impedance, but the capacity

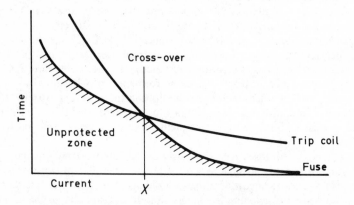

Figure 7.7 Back-up protection. If the current exceeds X the fuse must finish arcing before the trip mechanism moves; account being taken of the confidence limits (see also Figure 2.11)

should be checked. It is, however, very important when breakers are electrically close to an incoming supply or transformer.

The manner in which one protective device will back up another is illustrated in Figure 7.6. The take-over point should be on a part of the curve where the criterion becomes I^2 not $[I^2 t]$.

For higher current ratings the following values have been suggested:

Back up h.r.c. fuses — minimum probable $[I^2 t]$
Current rating (A)

Current rating (A)	$[I^2 t]$
60	20 000
100	80 000
150	180 000
200	320 000
300	720 000

which values might be above the withstand value of an mcb unless confirmed by tests (see earlier information on continental back-up fuses integral with miniature circuit breakers). The use of fuses to protect semi-conductor devices (rectifiers, transistors, thyristors, etc) are discussed in a later section.

7.14 Fuse design

It is not the purpose of this book to discuss the designer's job, but the following are some points which should be considered.

Electricity Regulation No. 5 requires that every fuse shall be so constructed and arranged as effectively to interrupt the current before it exceeds the working rate as to cause danger. It shall be so constructed, guarded and placed as to prevent danger from overheating or from the scattering of hot metal or other substances when it comes into operation. This is a fair criterion of fuse performance, which is met by most modern cartridge fuses so long as their rating and breaking capacity are adequate. In addition fuses must not cause dangerous voltage transients and it must be possible to replace them without danger of touching live metal.

Semi-enclosed fuses have one useful characteristic not shared by cartridge fuses; it is possible to see at a glance whether a blown fuse was caused by a short circuit, overload, or old age. In the former case there is always some blackening by metal oxide fumes and a large gap in the fuse wire with at least a few copper globules. An overload usually only causes a break in the middle with very few signs of arcing. Because of the low rupturing capacity, however, they are very rightly falling out of use.

It is often impossible to tell if a cartridge fuse has operated and, for this reason, larger sizes often have some form of automatic indication. It is useful to have available, already made up, a flash lamp battery and bulb and flexible leads such as is commonly used for checking continuity.

Because simple wire fuses may cause excessive voltage transients when they operate, a number of methods have been adopted for larger sizes to control the number of breaks and pre-determine their position. This and filling the cartridges with sand or other granular or powdered material which absorbs heat and metal vapour also serves to improve their operation and make special characteristics and closer protection possible.

As stated above, the dimensions of fuses should prevent the insertion of an incorrect fuse in any circuit, although this is only really effective with cartridge fuses; there are, however, difficulties. For example, all fuses for a 13 A fused plug must be of the same size and over fusing of small flexible cables is possible, but in practice this has not proved to be of great danger. Similarly, in distribution boards, fuseways will be provided to take fuses suitable for the outgoing cables, but the size may be too large for some of the connected apparatus, since it is impossible to be certain what the supply will be used for at a future date. It is therefore common practice to design fuses so that a small cartridge fuse can be fitted to a large fuse holder but not vice versa.

In the past, different h.r.c. fuse manufacturers have had their own standard dimensions, but this is now disappearing with increasing standardisation. It is important, however, that fuses of special design or different characteristics should not be interchangeable, as this can be dangerous.

Distribution boards should be designed to facilitate fixing labels to identify the fuses for particular services and also for a table stating the correct fuse for each fuseway. Where circuits are heavily loaded a fuse or distribution board should be closely associated with a switch which will break load current and close on the prospective short circuit in safety.

It is very important when purchasing cartridge fuses to insist on obtaining ones which have been certified by ASTA (Association of Short Circuit Testing Stations) or another responsible testing authority and are marked as complying with the appropriate British (or international) Standard.

7.15 Special fuse designs — characteristics and misuses

Some of the special purposes for which fuses are used are:
  (a) Time limited delay fuses are used to bridge the terminals of automatic trip coils, fed from current transformers on circuit breakers and controllers. These are usually single-wire fuses in glass tube cartridges; this allows a blown fuse to be quickly identified. The time delay fuse gives an inverse time grading

similar to that of a fuse or thermal trip, and fuses of different sizes give a range of fusing characteristics. The duty is not onerous.

(b) Heat or thermal fuses are sometimes used inside furnaces and ovens to limit the maximum temperature; in this case it is excess ambient temperature which melts the fuse, not excess current. Such fuses may be of noble metals (e.g. gold) to prevent oxidation.

(c) Instrument fuses may be used for a wide range of duties, e.g. to protect delicate components – such as moving coils against very small excess currents with low or high open circuit voltages.

(d) Telecommunication fuses have a very low rupturing capacity, e.g. 10 times load current. Thus a 10 A fuse would have a rupturing capacity of 100 A, and a 0.5 A fuse only 5 A.

(e) Motor car fuses. These are usually fairly large, e.g. 30 A, but the rupturing duty is undemanding. 12 V is insufficient to maintain an arc so that an immediate break occurs on melting.

Special fuses should have dimensions which prevent interchange. For example, on one occasion I found telecommunication fuses in the test prods of a medium voltage test instrument which should have had special fuses capable of dealing with both excessive voltage (such as accidental use on high voltage, e.g. 3000 V, terminals) and also a very high rupturing capacity. A rather long, small size h.r.c. fuse was required; a telecommunication fuse would almost certainly have exploded.

Potential trouble arose from the action of the 'services' in standardising the BS telecom fuse and using the same dimensions for fuses of high ruptiring capacity.

7.16 Differential protection

The whole question of differential protection apart from discrimination between overload devices and/or fuses has been left unexamined in this book. This is not because it is unimportant, but because there is extensive literature on the subject which is well-known to the professional engineers likely to be concerned.

7.17 Releases and relays

Releases, relays and fuses are the principal means of preventing damage to conductors and equipment by excess currents. They may be either electromagnetic or thermal. In the latter, an attempt is made to match the time/temperature characteristics of the protection with those of the

equipment protected, and the thermal element may be located within the protected equipment itself or in a circuit breaker or controller. It may be a pair of contacts operated by a bimetal strip, in which case it can be self resetting, or a heat or thermal fuse which blows when the ambient temperature rises as well as when the through current is excessive. Occasionally it may depend on a thermocouple or a resistance element in a balanced bridge network, but this requires special instrumentation. The sensing element operates a retaining or tripping coil in the control circuit on a relay.

Electromagnetic overload releases may be operated directly on the main circuit breaker or a contactor and may be instantaneous or have an inverse tune characteristic which should, ideally, match the thermal characteristics of the protected cables and apparatus. In some designs this characteristic of the setting (i.e. operating current) may be adjustable but is more effectively achieved by inverse time relays which are commonly used on public electricity supply networks and for major units in industry. It is possible to operate the release directly by the load current, but relays are commonly actuated from current transformers which may be located at the most sensitive point on the load; several c.t.s can be arranged to operate a single circuit breaker when this is desirable for discrimination.

The main advantage of thermal and magnetic overload releases is their ease of adjustment. Also, they are easier to reset than fuses, but this in itself can be a disadvantage if they are accessible to unskilled or irresponsible persons. The main disadvantage is that small circuit breakers and (nearly) all controllers have inadequate rupturing capacity for dealing with the heavy short circuits that may occur on some industrial installations (i.e. if close to large transformers) and they must then be backed up by high rupturing capacity fuses.

It is thus very important to ensure that overload protection is suitable for the apparatus and cables protected, that it has adequate rupturing capacity or is backed up by suitable h.r.c. fuses, and that it can provide adequate discrimination so that it always operates when required but does not do so unnecessarily. The designs of thermal and electromagnetic overload devices vary very greatly and a careful study must be made of the makers' specifications and instructions. The chapters on Earthing, Control, Fire Hazards and Detection (Chapters 8–12) are relevant to this matter.

7.18 Miniature circuit breakers

Because of their convenience, miniature circuit breakers are being increasingly used in place of fuses on distribution boards, but they have

limited short-circuit capacity and, particularly with the larger sizes, it is usually necessary to back them up by cartridge fuses. Matching the breaker and fuse characteristic is, however, difficult, since these breakers commonly have two release mechanisms, one for overloads and one for heavier faults; in addition, the settings may be altered, thereby changing the joint characteristic.

7.18.1 British Standard for m.c.b's

The appropriate British Standard is BS 3871, which lays down specific technical requirements. The usual form of the m.c.b. embodies total enclosure in a moulded insulating material. As the operating mechanism must be fitted with an automatic release independent of the closing mechanism, the m.c.b. is such that the user cannot alter the overcurrent setting nor close the breaker under fault conditions. At the same time the m.c.b. must tolerate harmless transient overloads while clearing short circuits. For most practical conditions, a changeover from time-delay switching to 'instantaneous' tripping at currents exceeding 6–10 times full-load rating is suitable.

Methods of achieving the required operating characteristics can be classified as (i) thermo-magnetic, (ii) assisted thermal and (iii) magneto-hydraulic. In the thermo-magnetic method the time-delay is provided by a bimetal element, the fast trip by a separate magnetically operated mechanism based on a trip coil. In the assisted thermal method the bimetal is itself subjected to magnetic force. The magneto-hydraulic mechanism incorporates a sealed dashpot with a fluid and a spring restraint, the dashpot plunger being of iron and subject to the magnetic pull of the trip coil.

7.18.2 Types of m.c.b.

There are two types of m.c.b., — high impedance and low impedance with different characteristics. When used in domestic and similar situations they usually perform well and give close protection (i.e. the ratio of the minimum operating current to the rating is low) but they will generally need back-up protection unless the rupturing duty is low and there are limits to its effectiveness. Integration into a distributive system with fuses may present problems if good discrimination and protection are to be combined. The subject is technically complicated and the review by H W Wolf[4] in 1970 should be consulted.

References

1. Stigant, S. Austen and Franklin, A.C., *The J & P Transformer Book*, 10th ed. Butterworths (1974)
2. Lythall, R.T., *The J & P Switchgear Book*, Butterworths (1972)
3. Lawson, C.S., 'Operation of fuses and MPB's on short circuit', *Elect. Times* (April 4 1957)
4. Wolff, H.W., 'Integration of m.c.b.'s into distribution networks', *IEE Review*, **117** (Aug 1970)
5. Lord Rayleigh 'Theory of Sound', *IEE Review* **2** (1970)

Chapter 8

Earthing Principles

List of symbols

V Voltage

I Current

R Resistance

r_{ab} r_{aa} Resistance coefficient; $r_{ab} = r_{ba}$ in normal circuits (without rectification) (these are analogues to the Maxwell potential coefficients in classical electrostatic theory)

$R_{ab} = r_{aa} - 2r_{ab} + r_{bb}$ = resistance between a and b

V_{ab} = *change* of voltage at 'a' caused by current I_b entering the system at b

   $= I_b r_{ab}$

ρ Resistivity

Note: For alternating currents we may substitute impedance for resistance coefficients (or operators for transients) so that

$$Z_{ab} = z_{aa} - 2z_{ab} + z_{bb}$$

This is, however, outside the scope of the present chapter. The advantage of the system is that we can solve problems in terms of inlets (or outlets) without considering the details of what lies between.

8.1 General principles

Earthing (or grounding as it is called in America) has played a central role in British practice but some aspects of the matter are often difficult to express for lack of a convenient notation. In this chapter I have described a system of analyses based by analogy on Maxwell's potential coefficients, which have been used before for calculating individual earth electrode resistances but not developed as a coherent system. Effective earthing may be considered under three headings:

(a) It prevents the outer casing of apparatus and conductors from assuming a potential which is dangerously different from the surroundings. Where there is an explosive risk, there may be a danger from very small voltages.

(b) It must allow sufficient current to pass safely to operate overcurrent or other protective devices, including those for earthleakage protection, without danger; this requirement may tend to conflict with (a) and thereby restrict the choice of methods of protection.

(c) It may be necessary to suppress dangerous earth-potential gradients. These may cause the incorrect operation of protective or control circuits and even destroy them, and also cause fire and explosion risks at places remote from the source of trouble.

The subject of earthing is covered by Code of Practice CP 1013[2]. This should be consulted for an explanation of this rather difficult subject.

8.2 The earth electrode problem, resistance coefficients

Standard network analysis relates to pairs of terminals, with current entering a network at one of a pair and leaving at the other. When dealing with 'earth plates' or electrodes only one 'terminal' is usually available. The other may be miles away or otherwise inaccessible. It may be only notional, the return path being by distributed leakage through cable or overhead line insulation over a wide area. On insulated a.c. systems it may be via distributed capacitance, in which case we must deal with earth loop impedance rather than earth loop resistance.

It has therefore become usual to speak of the resistance (or impedance) of single electrodes, but where two electrodes are fairly close together the resistance between them is *less* than the sum of the two earth resistances as measured by standard methods. This discrepancy is commonly called the proximity effect. On the other hand where several electrodes are bonded together to obtain a 'good earth' the joint electrode resistance is *greater* than that calculated for resistances in parallel. If there are for example four electrodes, there are six pairs and therefore six proximity effects to be considered.

In other circumstances we may need to calculate the potential gradients at the surface of the ground in the vicinity of a system of earth electrodes and/or fortuitous 'earths', following an accident or to determine the probable effects on instrumentation or protective gear. This usually involves determining the potential difference of points between which no current flows. Similar problems may arise when earth fault current follows a multitude of undetermined paths through the fabric of a building.

The problems are given precision by developing network theory in terms of single electrodes instead of pairs. If current enters (or leaves) a system at point *a* it will cause a *change* in the voltage between point *b* and some other 'datum' point. It also causes a change in the similarly measured voltages of points *c, d, e,* etc. But if we confine our attention to *changes* in voltage, these will be independent of the choice of datum point and we can ignore it.

The current must necessarily leave the system at some point, but unless this is a point in which we are interested we can ignore this also as it does not affect our calculations.

In this context, 'system' is to be interpreted very widely and includes for example earth continuity conductors, the frame and floors of buildings, water pipes and tramway rails, foundations (cement is a fairly good conductor for the present purpose) and 'the general mass of earth' (to adopt the term used in the Electricity Regulations). For this reason I have used a block diagram to represent a 'system' in Figure 8.1 instead of an equivalent circuit which does not indicate the underlying complexity.

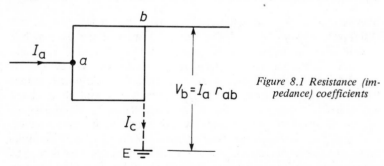

Figure 8.1 Resistance (impedance) coefficients

$$V_b = I_a \, r_{ab}$$

This approach, which is developed below, has the added advantage that, when dealing with earth electrode problems it is possible, by using the formal analogy with Maxwell's potential coefficients, to take over the results and formulae for systems of conductors in classical electrostatics, it being necessary only to introduce a simple numerical conversion constant.

8.2.1 Analytical development

If a current I_a flows to (or from) earth at a point a, either direct or through an earth-continuity conductor, the sheath of a cable, the frame of a building or by any other conducting path, there will be a change in potential of all the conductors connected to it. These are real and measurable voltages, and the *change in potential* V_b at any point b, whether on conducting metal or the surface of the earth, can be expressed as

$$V_b = I_a r_{ab} \; \Big\}$$

or $$r_{ab} = V_b / I_a \;$$ (8.1)

If the conditions were reversed, so that $V_a = I_b r_{ba}$ and r_{ab} is then equal to r_{ba}, unless there is an internal source of rectification. The change in voltage at a is, of course, $V_a = I_a r_{aa}$. These ratios are measured in ohms, but they are not resistances. It is convenient to call them resistance coefficients. In the particular example where a and b are two earth electrodes and current enters at a and leaves at b.

$$V_a = I_a r_{aa} + I_b r_{ba}$$

$$V_b = I_b r_{bb} + I_a r_{ab}$$

but $I_b = -I_a$ and $r_{ab} = r_{ba}$ in this problem

therefore $$R_{ab} = \frac{V_a - V_b}{I_a} = r_{aa} - 2r_{ab} + r_{bb}$$ (8.2)

It is convenient to use capital letters (e.g. R_{ab}) for resistances and small letters for resistance coefficients.

Where r_{aa} and r_{bb} are the resistance coefficients of earth electrodes, usually, but inaccurately, called the earth-plate resistances, we can recognise $2r_{ab}$ as the proximity effect, which expresses the reduction in resistance caused by the earth plates being within each other's area of influence. All these quantities can be measured and sometimes calculated (or estimated) and used independently of one another in dealing with earthing problems.

(There is a clear connection between such coefficients as r_{aa} r_{ab} and transfer a driving point impedance in standard network theory, but they are not identical and I have found it best to forget standard theory when dealing with earthing problems.)

By concentrating attention on the changes in potential and not bothering about the path of the current, which may well be diffuse or unknown, it is possible to state earthing problems more clearly and therefore to think about them more accurately which is illustrated by the following example.

8.2.2 Three and four terminal networks, Holmholtz theorem

If we are concerned with four terminals, and current I is led in at terminal 1 and out at 2 then the difference in potential which this causes between 3 and 4 is easily seen to be

$$V_{34} = (V_3 - V_4) = \{(r_{13} + r_{24}) - (r_{14} + r_{23})\}I \qquad (8.3)$$

$$= aI \quad \text{say}$$

Also if the same current is led in at 3 and out at 4 then the potential difference between 1 and 2 is

$$V_{12} = \{(r_{31} + r_{42}) - (r_{41} + r_{32})\}I$$

$$= \text{again } aI$$

This result is called Holmholtz' Reciprocal Theorem.

If there are only three terminals, e.g. points 2 and 4 are the same, the result is applicable to a 3 terminal network. It is sometimes convenient to write

$$a = {}^{12}r_{34} = {}^{34}r_{12} \qquad (8.4)$$

8.2.3 Potential gradient (Figure 8.2)

In earthing problems we frequently want to know the difference in potential between two points say 2 and 3 in the ground or on parts of a structure caused by current leakage to earth at point 1. This is obviously

$$V_{23} = I_1(r_{12} - r_{13})$$

The following simple theorem is also important. In Figure 8.2b, if all the current entering a conductor (or network) at A passes through B then

$$r_{ab} = r_{bb} \qquad (8.5)$$

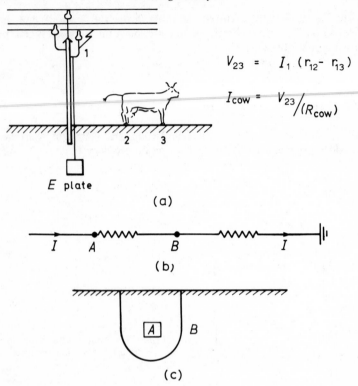

$$V_{23} = I_1 (r_{12} - r_{13})$$

$$I_{cow} = V_{23}/(R_{cow})$$

(a)

(b)

(c)

Figure 8.2 Potential gradients

This is easily proved as follows:

$$V_a = I_a(R_{ab} + r_{bb}) = I_a r_{aa}$$

in this instance, but

$$R_{ab} = r_{aa} - 2r_{ab} + r_{bb} \text{ generally}$$

Eliminating r_{aa} and R_{ab} gives $r_{ab} = r_{bb}$.

In Figure 8.2c, B is an equipotential surface surrounding a buried electrode A. It can be replaced by a conducting membrane without disturbing the flow of current, so that for any point on surface B,

$$r_{ab} = r_{bb}$$

This result may at first seem a little peculiar but its correctness is immediately apparent when it is observed that the contribution to the

potential at B by the current flowing through B from A does not depend on the resistance immediately around A but between B and 'the general mass of the earth'.

8.2.4 Earth electrode measurements (Figure 8.3)

The usual 'earth spike' method of measuring earth electrode 'resistance' provides a simple instance of the application of this notation

$$(V_e - V_1) = (I_e r_{ee} + I_1 r_{e1} + I_2 r_{e2})$$

$$- (I_1 r_{11} + I_e r_{e1} + I_2 r_{21})$$

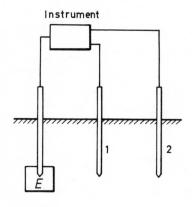

Instrument

Figure 8.3 Earth electrode measurement using earth spike method

The experiment is arranged so that $I_1 r_{11}$ is very small and, by putting the spikes well apart r_{e1}, r_{e2} and r_{12} are also small, so that

$$(V_e - V_1) = I_e r_{ee} \quad \text{very nearly} \tag{8.6}$$

even if r_{11} and r_{22} are quite large.

8.2.5 Earth resistivity (Figure 8.4)

In Wenner's test for earth resistivity, four spikes are driven into the ground at equal distances l apart. Current (I) is passed into spike 1 and

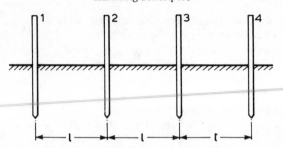

Figure 8.4 Wenner's test for resistivity

out from spike 4, the Holmholtz coefficient $^{14}r_{23}$ being measured. Then (taking account of the sign of I at 1 and 4)

$$^{14}r_{23} = \{r_{12} + r_{43} - r_{24} - r_{13}\}$$

but from symmetry

$$r_{12} = r_{34} \quad \text{and} \quad r_{13} = r_{24}$$

thus

$$^{14}r_{23} = 2\{r_{12} - r_{13}\}$$

Also, assuming the size of the spikes to be small compared with the distance between them (see section 8.5 Appendix).

$$r_{12} = \rho/2\pi l \quad \text{approx.}$$

$$r_{13} = \tfrac{1}{2}\rho/2\pi \quad \text{approx.}$$

Therefore

$$^{14}r_{23} = \rho/2\pi l \quad \text{or} \quad \rho = 2\pi l \,^{14}r_{23} \tag{8.7}$$

Where $^{12}r_{23}$ is the voltage between 4 and 3 divided by the current passing from 1 to 2.

8.2.6 Current picked up by well earthed cable sheath or rail, etc

If an earth electrode is near an extensive buried conductor such as a bare cable sheath, pipe, or rail, and the latter is well earthed because of its extent, then some of the current will enter this adjacent conductor.

Current going into ground from first electrode $= I_1$
Current picked up by nearby conductor $= I_2$
Voltage applied to electrode $= V_1 = I_1 r_{11} + I_2 r_{12}$
Voltage rise of second conductor $= V_2 = I_2 r_{22} + I_1 r_{21} = 0$ approx.
Ratio of currents $I_2 = -I_1 r_{21}/r_{22}$
Reduction of effective earthing resistance $V_1/I_1 = (r_{11} - r_{21}^2/r_{22})$
Current picked up in terms of volts supplied
$$I_2 = V_1 r_{21}/(r_{21}^2 - r_{11} r_{22})$$

8.3 Earthing of installations and apparatus

8.3.1 Earth electrodes

In some installations, it is extremely difficult, if not impossible, to 'connect to the general mass of earth', as required by the regulations. On one occasion, in an ironstaone area, a disused steel artesian well casing going down several hundred feet was used, but, when it was measured, its resistance coefficient was found to be several hundred ohms, therefore the alternative of other suitable means of preventing danger must be adopted; the practical goal being to prevent accessible metal from reaching a potential dangerously different from its surroundings.

Steel-framed industrial buildings are generally well earthed, the footings of the stanchions being, in effect, a number of earth electrodes in parallel. Otherwise, a number of driven rods connected in parallel are often effective (6ft rods, 1 in diameter and 6 ft apart usually form a satisfactory arrangement). In shallow soil over rock, horizontal earth strips or a mesh may be effective, but all earth electrodes must be tested before they are accepted. There are standard and efficient commercial test sets for this purpose.

Methods of measuring earth electrode resistance coefficients are given below and calculations in the appendix to this chapter.

IEE Regulations do not permit the structure to be used as the sole earth-continuity conductor. This is probably wise, since paint and rust may insulate members, but the actual footings bonded in parallel may well be better than any specially installed electrode. This must, however, be checked by tests.

The recommendations of the Electricity Council, dated March 1962, on earthing consumers' installations are given in CP1013: 1965, Clause 607, and area boards commonly permit the use of their cable sheaths and multiple earthing conductors etc for this purpose, but, as a condition of supply, they accept no responsibility for the arrangements.

At one time, it was common practice to use a water main as an earth electrode, although this was not welcomed by the water authorities; but, now that nonmetallic water pipes are increasingly used, this practice is undesirable.

8.3.2 Earth-loop impedances (Figures 8.1 and 8.5)

The current that will flow on the occurrence of an earth fault is equal to the line-earth voltage divided by the earth-loop impedance. In most situations, the transformer and fault are not close, and $R_{12} = r_{11} + r_{22}$.

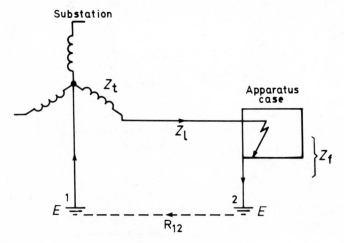

Figure 8.5 Loop impedance
$$I = V/(Z_t + Z_p + Z_f + R_{12}) = V_1/Z_{loop}$$
l_1 *indicates earth fault*

The earth-loop impedance is almost impossible to calculate accurately and should be measured; but, because it contains an unpredictable part, i.e. the earth-fault impedance Z_f, a considerable safety factor must be used in its application. When the protection is by fuses, which is common on small installations or on minor circuits in large installations, HM Inspectors have recommended a factor of 4, i.e. a safety (ignorance) factor of 2 and an allowance of 2 for the fusing factor, in selecting the fuse size (i.e. normal current rating of fuses), so that the maximum size of fuse is

$$F = \frac{\text{voltage to earth}}{4 \times Z \text{ loop}} \text{ amps} \tag{8.8}$$

A smaller fusing factor is available for cartridge fuses, but it is best to use this to increase safety, particularly in the situation shown in Figure 8.6 where there is no metallic return path.

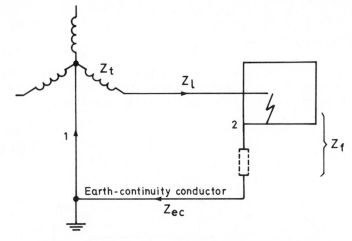

Figure 8.6 Loop impedance with earth continuity conductor

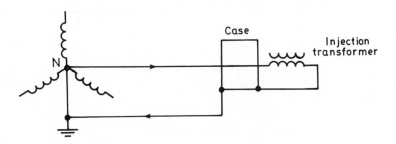

Figure 8.7 Measurement of neutral-earth loop impedance

There has been considerable controversy about the measurement of earth-loop impedance; much of it ill-conceived. Strictly, the impedance of the actual fault path, as shown in Figure 8.6, should be measured, but this entails either making the tests when the plant is shut down or working on live circuits. If a simulated fault is used and the metallic return path happens to be broken or the earth resistance R_{12} is unexpectedly high, the result may be that, momentarily, the potential of accessible metal, such as conduit or motor starters, is raised to the line voltage. The risk may not be very great, but a safety procedure should not deliberately introduce the possibility of a serious accident.

For this reason the late H. Midgley and I suggested that the error introduced by measuring the neutral/earth-loop impedance (Figure 8.7) instead would not be important in the routine testing of installations, and any error would overestimate the impedance and give increased safety, since the neutral conductor is usually smaller than the line conductor. It was not intended that it should be very accurate; an error of 10 or 20 % is immaterial, since the eath-loop impedance is not a fixed quantity but may fluctuate from day to day, depending on weather conditions and other fortuitous influences.

8.3.3 Earth continuity

Whatever method of earthing is adopted, the continuity of the earth conductor is of vital importance. In heavy industry, apparatus is often

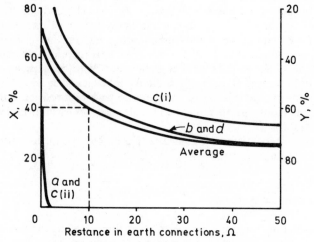

Figure 8.8 Continuity tests on factory earthing systems
X = *percentage of installations examined having resistance above the value shown by the curves*
Y = *percentage of installations examined having resistance below the value shown by the curves*
(a) Continuous screwed conduit installation in good condition in steel frame and galvanised-iron building
(b) Conduit in brick or steel and timber buildings and conduit with sections bonded by earth wires and clamps
(c) Conduit and straying earth wires
(i) Old brick buildings with wooden floors and conduit or straying earth wires: normal conditions
(ii) As (i) but earthing overhauled following fatal accident
(d) Miscellaneous; in general, heavy-engineering works, frame of building, straying earth wires etc

bolted to stanchions or other parts of the structure and is most probably earthed independently of other provisions. In brick or timber buildings, however, there is not the same safety net. Screwed conduit in good condition forms a good earth path, but ERA reports have shown that, with age, it may well become unreliable owing to rust in the joints and slackening off of the connections. For this reason, all joints should be pulled up tight with back nuts, and these are particularly important at the entrance to the apparatus. At one time, it was common to see the conduit stopping short and being bonded to the apparatus casing by a twiddle of wire. This was very unreliable. Metal sheaths and/or armour should terminate in properly constructed glands securely attached to the apparatus. The importance of these matters is made clear by Figure 8.8.

Generally speaking, every advantage should be taken of bonding to independent parallel earth paths, such as structural steel, and to works water mains, when these are known to be well earthed. Isolated machines in the centre of a shop fed from overhead busbars are usually earthed by connection to the busbar casing, but this is vulnerable to shocks and vibration and other mischances. It is therefore advisable to have an auxiliary earthing path at ground level, with, perhaps, a copper strap. For mechanical reasons, this should not be less than, say, $1 \times 1/8$ in. Flexible tubing is liable to crack and to have high-resistance terminations; it should not therefore be the sole means of earthing motors or apparatus.

Very good commercial combined earth-electrode and earth-continuity testing sets are available.

8.3.4 Substation earthing

At a substation, there are normally at least three earthing requirements: for the casing of incoming high-voltage switchgear and transformers, for the neutral conductor of the outgoing medium-voltage supply and for the casing of medium-voltage equipment. At one time, it was common to bond all these together, but the following incidents illustrate the dangers which may arise.

In Figure 8.9a the fault current is $I = V/$h.v.-earth-loop-impedance, and the rise in voltage of the transformer casing is Ir_{22}. It was calculated that Ir_{22} was approximately 4500 V, and the result was that this was communicated via the medium-voltage neutral conductor to the outgoing distribution cables, and it blew up the supply meters in a row of houses and caused a fire, which was fortunately not a serious one.

Similarly, on a farm (see Figure 8.9b), when a high-voltage high-resistance fault on a pole transformer occurred, it caused a difference in

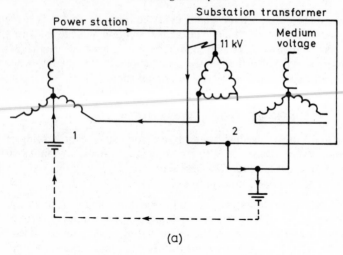

(a)

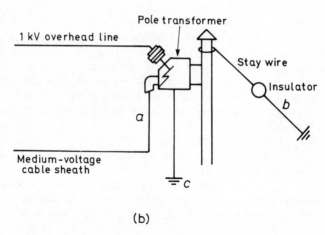

(b)

Figure 8.9(a) Dangers involved in substation earthing
(b) Substation-earthing dangers. A typical case where a high-
voltage high-resistance fault occurred on a pole transformer

potential between the outgoing medium-voltage cable sheath and the
lower half of the stay wire to the pole, and a man who happened to
touch both was killed. The shock voltage would be $(V/Z_{loop})(r_{ca} - r_{cb})$.
CP 1013 may be consulted on this point.

There are a number of possible permutations of such circumstances.
The general conclusion is that all accessible metal should be bonded
together, but earth electrodes which may rise in voltage substantially

when a fault occurs should be remote from other internal or fortuitous electrodes and connections. The actual solution will depend on the individual circumstances, but one useful expedient on overhead-line supplies is to earth the neutral conductor of the medium-voltage system one span length away from the substation.

8.3.5 Potential gradients and intrusive potentials

Among the most insidious hazards are conductors which bridge the distance between areas at different potentials, owing to faults, earth leakage or stray currents. The best way of dealing with this problem is probably to describe a number of examples.

(*a*) A power station and a grid substation, about 200 yd apart, were connected by both power and control cables. A heavy earth fault in a grid transformer caused the associated metalwork to be raised by, probably, 2000 or 3000 V. The sheaths of the cables were interrupted, to prevent circulating currents in the event of faults, but this established a voltage across the break in the sheath of a control cable which caused an arcover, and the cable end was destroyed and other damage done, so that control was lost. In my recollection, the result was that the running generator had to be shut down using the stop valve.

(*b*) A fault on an electric radiator in a garage office caused sparking at a casual contact in a cellar some distance away. This ignited vapour from petrol, which had seeped through from a storage tank, and caused an explosion.

(*c*) A shock was received from a post and wire fence, the other end of which was in contact with metalwork in the vicinity of an overhead-line pole, when there was an uncleared fault.

(*d*) Vapour ignition and fire at one part of a works was caused by arc welding in a yard some hundreds of feet away, the welder not having bothered to provide an adequate return path for the welding current.

These and many other incidents emphasise the importance of providing adequate bonding and conductivity to suppress any secondary pressure rise. Apart from shock risks, there are particular hazards where there is a serious fire or explosion risk, and also earth potential gradients may seriously disturb, or even destroy, telecommunication, instrumentation and control equipment.

8.3.6 Impedance earthing and earth-leakage circuit breakers

When natural earthing is very good or there is a high-conductivity earth conductor, very heavy earth-return currents may flow, and this may

lead to damage to apparatus before protective equipment operates or to fire risks. If the fault impedance forms the major part of the earth-loop impedance, there may be a dangerous local voltage rise, leading to hazards of the type described above. In such circumstances, an impedance may be inserted between the main transformer neutral conductor and the earth (or the earth-return conductor). This is not often adopted, however, on low- and medium-pressure supplies unless the transformers are very large. (If the reactance of the connection between the transformer neutral conductor and the earth is tuned with the effective system capacitance with one line earthed it is called an arc-suppression coil; this is outside the scope of this chapter, but see section 6.14).

Earth-leakage protection is of two types, depending on either the detection of the out-of-balance current or the consequent difference of voltage between the metal casing or sheaths of apparatus and conductors and adjacent metalwork, concrete floors, damp earth, window frames, steam pipes, structural steel etc., since concrete and building materials normally conduct sufficient current to permit a fatal electric shock.

The relationship between the earth-loop impedance and the rating of fuses or the overload settings of switches and circuit breakers, when these alone provide the protection, has been discussed above, but the restrictions which their use introduces have led to the development of special small circuit breakers with more sensitive detection. For many years, earth leakage on major power circuits has been effected by the use of summation-current transformers; so that, unless the vector sum of the phase and neutral currents is zero, a current or voltage is induced in the secondary circuit which is proportional to the out-of-balance or leakage current and is used to trip a circuit breaker. In some modern small circuit breakers, there is not a separate current transformer, but one magnetic circuit combines the functions of out-of-balance detection and operation. Such devices can be made very sensitive, but there is a lower limit to the out-of-balance current which can be used, since, if it is too small, the setting may be less than the capacitance and normal leakage current of the circuits and apparatus protected. 0.5 A has been suggested, but this by itself is not adequate for protection against electric shock. The IEE Regulations require that the product of the earth-loop impedance and the operating out-of-balance current shall not exceed 40 V (Regulation D24). This limits the sustained voltage but not the momentary rise, and it assumes that the loop impedance is known and constant, or at least has an upper limit. In many situations, it may vary considerably, but it is not easy to suggest a better criterion.

Voltage-operated circuit breakers detect the difference in potential

between the casing of protected apparatus, or its earth-continuity conductor, and an earth electrode at a suitable point not in the immediate vicinity (Figure 8.10). This is usually set at not more than 40 V. The condition for operation is then $V = I_f(r_{fa} - r_{fb}) > 40$, where $I_f =$ fault current.

This can give protection against fatal electric shock, but not against damage or fire risk from overcurrent. It also has the disadvantages for industrial installations that it may be difficult to ensure that the main

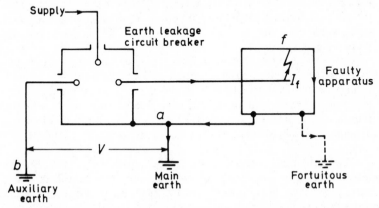

Figure 8.10 Application of voltage-operated earth-leakage circuit breaker

and auxiliary earth electrodes are sufficiently separated; i.e. $r_{fa} - r_{fb}$ is sufficiently large. It is also possible for them to become fortuitously bridged by an unnoticed conductor, or a better-earthed conductor from outside may intrude into the protected areas, so that the voltage exceeds 40 V.

Finally, we have the following distinction: out-of-balance protection can be applied to a particular cable or apparatus, and the area of disconnection is limited in the event of a fault; whereas voltage-operated protection may give better protection against shock, but can only be applied to a location from which the whole of the supply must be cut off. For this reason, the two forms of protection may be combined with advantage.

8.3.7 Earthing to neutral — concentric cables and continental practice

In Europe (except the U.K.) it has been a common practice to use the neutral conductor of a medium-pressure 3-phase system as the earth-continuity conductor. For example, imported machine tools may have contactor coils connected between the phase conductors and the frame

of the machine, and the same applied to bracket lamps. This is contrary to the United Kingdom Electricity Supply Regulations, which state that, except with the approval of the Ministry, a system may be earthed at one point only. The Regulations do not apply within a factory supplied from its own transformers or generators, but this is considered to be bad practice and potentially dangerous. If there is an accidental break in the neutral conductor, it may result in machines over a wide area becoming live at line voltage.

However, the Factory Electricity Regulations have always permitted concentric cables with the outer conductor earthed and exposed (see Regulations 9 and 10), but no switch, fuse or link may be placed in that conductor, except for testing at the power station (the Regulations should be consulted for the exact wording and requirements). This provision was included because, in the early part of the 20th century, proprietary concentric systems were in use, but the practice was obsolete and almost entirely abandoned until relatively recently, when it was revived with the introduction of single-core metal- (copper or aluminium) sheathed mineral-insulated (m.s.m.i.) cables.

The safety of such systems is entirely dependent on the effective continuity of the neutral conductor. So far, this has not caused much trouble in steel-framed industrial buildings, where there are many fortuitous parallel earth paths, or in buildings such as blocks of flats, where the danger of damage or corrosion is low, and a high standard of installation work under close supervision is possible. However, it is inexcusable where there is a serious fire or explosion hazard, and there is considerable doubt as to its safety in other buildings where standards may not be so good.

The copper sheath may work-harden and crack where it is subject to vibration or repeated flexing, such as at the input to motors or on travelling-crane structures. Copper has been known to corrode so seriously that unprotected m.s.m.i. cable has had to be abandoned on some coke ovens. Copper will also corrode in sea air; generally speaking, it does not resist combined acid and oxidising conditions.

8.3.8 Protective multiple earthing of neutral conductor

Where local earthing is difficult and particularly where the supply is via overhead lines, it has been common practice for the line to carry both an earth conductor and a neutral conductor insulated from one another; but, in recent years, consumers have been allowed to earth the casing of their apparatus by connection to the neutral conductor, steps being taken to ensure that the resistance to earth is never greater than 10 ohms. The neutral conductor is earthed both at, or near, the transformer and

the far end of each line or branchline and at regular intermediate points, to reduce the danger of a voltage rise in the event of a fracture of the neutral conductor. (This matter is under discussion, and in the U.K. the Electricity Division of the Department of Energy should be consulted.)

The Factory Inspectorate recommend that where this system has been adopted by the supply authority, there should be as good an earth as possible near the supply intake, and that, from this point onwards, the installation should be normal, with separate insulated

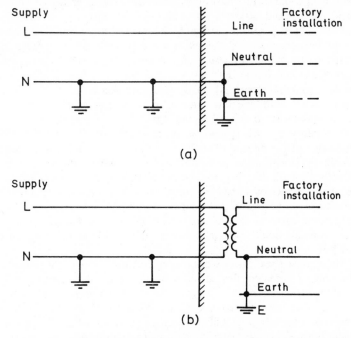

Figure 8.11 Supply to industrial premises from system with multiple earthed neutral
(a) with separate insulated neutral and earth continuity
(b) with isolating transformer

neutral and earth continuity for enclosures and cable sheaths (Figure 8.11a and b). There is an advantage, however, in adopting the arrangement in Figure 8.11b, since this ensures that, in the event of the neutral conductor breaking on the supply side of the isolating transformer, there is a closed path for both load (LN) and fault currents (LE) which will tend to prevent any rise in voltage due to returning load currents, and operate protection in the event of a fault. This scheme can be

backed up by voltage-operated protection between E and an auxiliary electrode.

8.3.9 Other earthing systems

Earths may also be required for lightning protection, the dissipation of static charges in danger areas and telecommunication, such as telephones or radio and television. (This does not necessarily apply to the small domestic type of radio and television sets, which are usually in insulating cases.) In general, these should have separate earth electrodes; so that a heavy current in one system (including the power-supply earthing) does not induce dangerous currents and voltages in the others.

8.3.10 Transportables

If transportable apparatus is to be earthed, for reasons of safety, it is most important that the earth continuity conductor in the flexible cable shall not break. If it does break or become disconnected, this should cause immediate failure to safety. For transportable equipment, this may be achieved by circulating a small current at a low voltage (e.g. 12 V or less) through two earth conductors in parallel and using this to energise the hold-in coil of a contactor (Figure 8.12). This system has

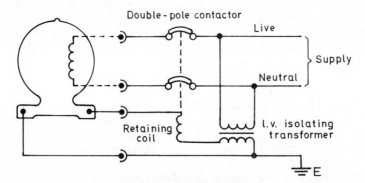

Figure 8.12 Basic monitoring circuit for earth continuity. (By permission of the Controller of Her Majesty's Stationery Office)

been in use in mines for a long time. If there is a short circuit between the two conductors, this simple design does not fail to safety. However, there are a number of ways in which this may be overcome, and circulating equipment, or monitored earthing, may be combined with

straightforward current balance or voltage-operated earth-leakage protection. The point which must be considered in a particular example is how far the increased protection is counterbalanced by the difficulties of effective maintenance, due to the added complication. A substantial number of permutations and combinations have been proposed and used. The application of such schemes was unfortunately hampered for a long time by the absence of a suitable British Standard for polyphase industrial plugs and sockets, but this gap has now been filled.

However, a very substantial degree of safety may be obtained by the use of reduced supply voltages (e.g. 110 V with the neutral or midpoint earthed) supplied via safety isolating transformers. This scheme also allows any earth leakage to be returned directly to the transformer neutral or midpoint via an earth-leakage trip. Since this is connected to only one (or, at the most, a small group) of machines, the setting may be much lower than would otherwise be practicable. Although some people think that 100 V with the neutral or midpoint earthed is too low for the larger transportable machines, this is not necessarily true, since, for many years, 50 or 100 ton steelworks cranes operated on 100 V supplies without difficulty.

8.3.11 Portable handtools

The same precaution may, in principle, be applied to portable hand-tools, but, in practice, the monitored-earth system becomes too complicated, unless a tool is always supplied from the same fixed point. For this reason, three types of protection against shock have been developed:

Class I: ordinary tools which require earthing
Class IIA: all-insulated tools
Class IIB: double-insulated tools
Class III: tools operating on extra low voltages; i.e. not exceeding 50 V (direct or alternating), with a maximum voltage to earth not exceeding 30 V (alternating) or 50 V (direct).

Class III tools for a.c. circuits should therefore be supplied from safety isolating transformers with the midpoint or neutral earthed (hence the 30 V maximum) for alternating current.

(*Note:* The definition of classes is that used in BS 2769: 1964 'Portable motor-operated power tools' and discussed in Chapter 4. Some continental practice does not distinguish between classes IIA and IIB, but double-insulation is specifically dealt with for the UK in the statutory Electricity Regulations and in the IEE Regulations.

Double insulation means that, between any exposed or accessible metalwork and live conductors, there are two layers of insulation: functional insulation, which surrounds the conductors, and protective insulation overall. The functional insulation can be an air gap or creepage path over insulation, the details of which are prescribed. Outer casings etc. need not all be of insulating material, provided that they are separated from live parts by double insulation. For example, for the chuck of a drill, the protective insulation may be an insulating sleeve between the armature and the spindle or, better still, a nylon gear between the spindle and chuck. Very efficient steps in detailed design must be adopted, and provision must be made to prevent a tool being incorrectly reassembled after adjustment or repair.

The constructional precautions necessary for safety have proved surprisingly difficult to prescribe in detail, and reference should be made to

BS 2769: Portable electric motor-operated tools
BS 3535: Safety isolating transformers
BS 2754: Memorandum on double-insulated and all-insulated electrical equipment

The problems arising in the design and use of double-insulated equipment are discussed in Chapter 7.

8.3.12 Transportable generating sets

On construction sites and for some other purposes, it may be necessary to use temporary transportable generating sets for lighting or tools. The reduced voltages discussed above should be used, and the earth-continuity conductors, which should be insulated and not generally accessible, should be brought directly to the neutral conductor or mid-point of the generator, or to one terminal if very low voltages are used. It is then possible to provide simply and efficiently current-operated earth-leakage protection combined with very sensitive voltage-operated protection, to hold the earth-continuity conductors down to the potential of the surroundings. This is discussed in CP 1013: 'Earthing'.

8.4 Earth electrode calculations (detailed considerations)

It is possible to use the close correlation between Maxwells electrostatic coefficients and earth resistance coefficients to obtain formulae for calculating the latter. This is explained in the Appendix to this chapter

and a brief table of values is given. This procedure has in effect been adopted by a number of writers in the past, but without taking full advantage of the facility it provides for explicitly separating the so-called 'earth plate resistance' from the proximity effect as two independent quantities.

Calculated coefficients are, however, an uncertain guide because the indeterminate nature of the resistivity of the earth, which may in fact vary from day to day and may be increased by 'drying out' under heavy fault conditions. For this reason they should always be checked by measurement. This aspect has been examined by a number of writers (see bibliography).

The analytical approach has, however, considerable value in comparing the effect of various types of earth electrode and methods of installation — such as depth of burying and distance apart, which are discussed in the following sections.

One theoretical point should be noted: electrode resistances and resistance coefficients vary directly as length so long and only so long as the total situations are geometrically similar. Thus the coefficient for a buried electrode will be twice that of a similar one twice as long buried at twice the depth, but not as one twice as long at the same depth. A proper understanding of this point resolves some of the apparent discrepancies which have been noted in the past.

8.4.1 Reflections or images

The use of 'reflections' is as useful here as in electrostatics and has a wider field since we can have virtual images in insulators as well as conductors. On the other hand the elaborate calculations made in electrostatic theory would be entirely unjustified because of the inherent uncertainty of the value of the resistivity ρ in earth electrode problems.

If two conductors are arranged symmetrically about an *imaginary* plane as in Figure 8.13, then if current +I enters at 1 and −I at 2 the change in potential at *any* point 3 on the plane is

$$V = I(r_{13} - r_{23})\qquad\text{(8.5 repeated)}$$

but from symmetry $r_{13} = r_{23}$ and $V = 0$; therefore the plane is an equipotential surface and a perfectly conducting sheet could be placed in the plane without disturbing the currents or potentials. That is to say if a source of current is near an extended conducting plane the potentials are the same as those which would arise if there were no plane but a source of equal and opposite magnitude at the same distance behind the plane. A conducting plane may therefore be called a *negative* mirror.

With the same arrangement but two sources of the same sign (e.g. $+I$ and $+I$) it is easy to see that no current would flow across the median plane, which could therefore be replaced by a perfectly insulating sheet. An insulating surface may therefore be called a *positive* mirror.

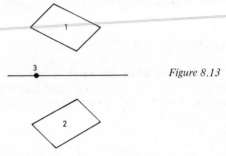

Figure 8.13

If finally the median plane separates two masses of resistivity ρ_1 and ρ_2 and there is a source at 1 then we may obtain the current and voltage pattern in mass 1 by assuming that there is a fictitious source at 2 and that

$$I_2 = I_1 \frac{\rho_2 - \rho_1}{\rho_2 + \rho_1} = KI_1 \quad \text{(say)}. \tag{8.9}$$

It will be noted that

$$\begin{aligned}
&\text{if } \rho_2 = \infty \text{ (insulator)} &&I_2 = +I_1 \\
&\text{if } \rho_2 = -\infty \text{ (conductor)} &&I_2 = -I_1 \\
&\text{if } \rho_2 = \rho_1 &&I_2 = 0
\end{aligned}$$

which agrees with the statements above.

8.4.2 Effect of depth of burying of earth electrodes

The obvious application of a positive mirror is to a buried earth plate, the surface of the ground being, in effect, a perfect insulator, then

$$V_1 = I_1(r_{11} + r_{12})$$

$$\text{i.e., } r'_{11} = (r_{11} + r_{12})$$

For a square earth plate (see Appendix 8.5)

$$r_{11} = \rho/4.5l$$
when $l = $ side in cm.
$$r_{12} = 0.9\rho/8\pi d \tag{8.10}$$

where d = ½ distance between images. 0.9 being estimated correction factor as in section 8.4.3 below.

These values are approximately the same as those given by Higgs more complicated formulae the difference being well below the probable errors due to variability of ρ.

8.4.3 Stratified earth

Equation 8.9 can be used when the earth is stratified with abrupt changes of resistance but it must be remembered that the earth plate is between two mirrors (Figure 8.14) and that strictly speaking there is an

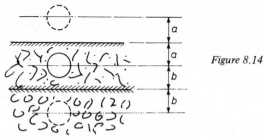

Figure 8.14

infinite regress of reflections as shewn in some books on light. Generally speaking however, only the primary reflections will be sufficiently close to have much effect, thus

$$V_1 = I(r_{11} + r_{21} + Kr_{31}) \qquad (8.11)$$

to the first order of reflections. K = estimated correction required if a or b is small compared with size of earth plate. This was taken as 0.9 in the previous section.

8.4.4 Other examples

Figures 8.15 and 8.16 illustrate the effects of an old, virtually insulating, foundation wall and an earth plate suspended in a dock near a ship. The manner in which calculations should be made is obvious.

8.4.5 Conclusion

The principal advantage of the notation suggested is that it enables one to state many problems clearly and concisely thus avoiding the rather

verbose and confusing arguments which are often used. Some results are difficult to state in any other way.

Earth plate calculations can never be very accurate and elaborate analytical formulae are not justified; their principal use is in discussing

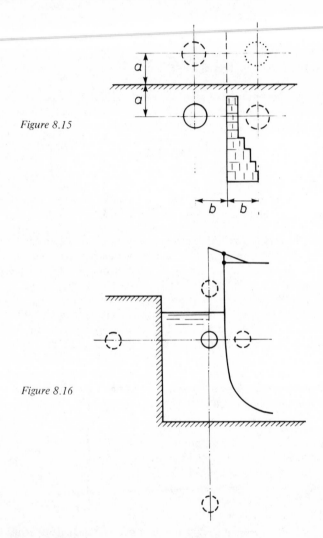

Figure 8.15

Figure 8.16

the relative merits of different procedure and it is not important therefore that the earthing coefficients are sometimes difficult to measure or calculate very accurately.

8.5 Appendix

There is a simple relation between CGS electrostatic formulae and resistance coefficients, i.e.

$$r_{nm} = \frac{\rho}{4\pi} p_{nm}$$

where ρ is in ohm cms;
p_{nm} is the Maxwell potential coefficient.

Thus, Sir James Jeans in *The Mathematical Theory of Electricity and Magnetism* gives the resistance between two perfectly conducting spheres immersed in a conducting mass of resistivity ρ as

$$R = \frac{\rho}{4\pi} (p_{11} - 2p_{12} + p_{22})$$

This corresponds to our

$$R = (r_{11} - 2r_{12} + r_{22})$$

$$\left.\vphantom{\begin{array}{c}a\\a\\a\\a\end{array}}\right\} \quad (8.12)$$

It is best to retain ρ *in ohm cms and dimensions in cms* to avoid mistakes in adapting formulae for potential coefficients from classical works on electrical theory, so as to avoid errors. A small selection of useful formulae is given below for the self coefficients r_{nn}.

For pairs of electrodes it is usually sufficient to take r_{nm} as $\rho/4\pi a$ where a is the distance apart of their centres, unless they are quite close together. This is particularly useful in calculating the effect of depth of burying by taking the joint effect of the electrode and its image.

Useful formulae for earth electrode resistance coefficients r_{nn}
(ρ is in ohm cms and dimensions are in cms)

Circular disc	$\rho/8a$	
Squares	$\rho/4.5l$ approx.	
Rectangle	$\rho K(e)/7.1l$ approx.	(8.13)
Narrow strip	$\rho\{ \log h(4l/d)\}/4l\sqrt{\pi}$	
Long rod	$\rho\{ \log h(2l/d)\}/2\pi l$	

where a = radius;
        l = side of square and long side of rectangle or strip;
        d = width
        $K(e)$ is the complete elliptic integral given below;
        $e = 1 - (d/l)^2$

e^2	$K(e)$	e^2	$K(e)$
0.0	1.571	0.5	1.854
0.1	1.612	0.6	1.950
0.2	1.660	0.7	2.075
0.3	1.714	0.8	2.257
0.4	1.778	0.9	2.578

The accuracy of the approximation that in many situations

$$r_{nm} = \rho/4\pi d$$

may be judged from the following table based on calculations by Lord Kelvin for two spheres of unit radius and distanced between centres:

d	$1/d$	$p_{12} = (4\pi/\rho)r$
2	0.500	0.721 (touching)
2.1	0.476	0.509 (very close)
2.5	0.400	0.406
3.0	0.333	0.335
4.0	0.250	0.250

If desired, this may be used as a basis for estimating corrections to the simple formulae.

Coefficients can be calculated accurately for only a few simple shapes, notably ellipsoids (Figure 8.17) but the results may be used to obtain approximate values for other shapes. [An ellipse may be defined as a closed curve which is the locus of all points for which the sum of the distance from two fixed points (the focii) is equal to the major axis.] For example, the coefficients for a rectangle (including narrow strips) may be taken as equal to those for elliptic discs of the same area and the same ratio of length to breadth. For a square this gives a value quite close to that which Maxwell obtained by rather tedious numerical integration and closer than other approximations commonly used. The coefficients for rods are close to those for long (needle-like) ellipsoids of the same length and diameter. This is the basis of the formulae given above (Figure 8.18).

The equipotentials around (deeply buried) ellipsoids are also ellipsoids, with the same focii but greater lengths. The equipotential surfaces are equivalent to perfectly conducting surfaces for which the mutual coefficients are calculated by $r_{ab} = r_{bb}$ (section 8.23). At

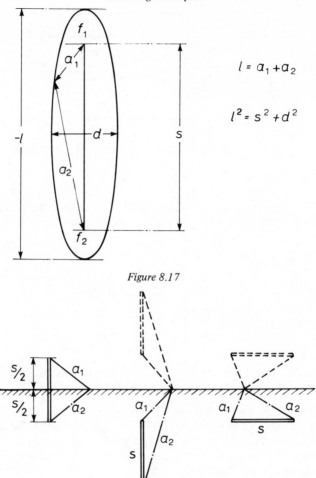

$$l = a_1 + a_2$$

$$l^2 = s^2 + d^2$$

Figure 8.17

Figure 8.18

comparatively short distances they approach the form of spheres so that

$$r_{nm} = \rho/4\pi a$$

where a is the radius of a sphere and is equal to the distance apart.

The best source of information on the electrical properties of ellipsoids from which closer approximations can be deduced, is probably Routh's *Analytical Statics*, (Vol. 2); Jean's *Mathematical Theory of Electricity and Magnetism* is also useful.

Other formulae based on different approximations are sometimes used. Though they look very different the numerical values obtained are often quite close.

Bibliography and References

1. Attwood, C., *Electric and magnetic fields*, John Wiley
2. CP 1013: 1965, *Earthing*, British Standards Institution
3. *Copper for earthing*, The Copper Development Association
4. *Earth Resistance Measurement*, Evershed & Vignoles Ltd
5. Elliott, N., 'Earth leakage protection in parallel with a solid earth', *Proc IEE* (June 1952)
6. Emerson, S.J., 'Protective multiple earthing', *Elect Supervisor* (Feb 1957)
7. Fawcett, S. and others, 'Practical aspects of earthing', *Proc IEE* (1939/40)
8. Gosland, L., 'Cost and efficiency of earthing low and medium voltage o.h. line systems', *Proc IEE* (1950)
9. Jahnke, E. and Ernde, F., *Tables of functions with formulae and curves*, Dover Publications, New York (1945)
10. Jeans, Sir James, *The mathematical theory of electricity and magnetism*, Cambridge University Press (1925)
11. Mann, F.H., 'Earthing in practice', *Elec Times* (March/April 1954)
12. Mather, F., 'Earthing low and medium voltage systems', *Proc IEE* (1958)
13. Maxwell, J.C., (ed) *Electrical researches of Henry Cavendish*, Lib of Sci. Classics, F. Cass (1967)
14. Morgan, P.D. and Taylor, H.G., 'The resistance of earth electrodes', *ERA Report F/T 50* (1932)
15. Routh, J.I., *Analytical statics*, Vol 2, Cambridge University Press
16. Tagg, G.F., 'Measurement of earth resistances', *Elec Times* (April 1931)
17. Tagg, G.F., 'Earth resistivity measurements', *Elec Times* (Sept 1935)
18. Tagg, G.F., 'Measurement of earth-electrode resistances – systems covering a large area', *Proc IEE* (1964)
19. Tomkins, A.H.E., 'Supply network earths', *Elec Times* (May 1935)

Chapter 9

Switches and Control Systems

9.1 Introduction

This chapter refers mainly to conventional controls and explains fundamental principles. A large number of conventional systems are still in use and will probably never be entirely eliminated. Modern solid state devices are also discussed and their intrinsic characteristics examined, but developments in this field have been so rapid that it is not possible to deal with particular devices in detail without the risk (almost the certainty) of the discussion quickly becoming out of date.

Overload and over voltage protection, and circuit breaker and fuse characteristics are mentioned only incidentally. These have already been discussed more fully in Chapter 6.

9.2 High voltage equipment

For many years isolators or isolating switches have been used with high voltage (1000 V and upwards in this context) switchgear. British Standards have defined an isolator as 'a device used to open or close a circuit either when negligible current is interrupted or established or when no significant change in the voltage across the terminals of each pole will result from the operation' e.g., when one of two parallel isolators is opened and continuity is maintained through the other.

This is a valuable definition as it draws attention to the limitations of a plain break or similar isolator under present-day conditions. In all other cases a switch should be used, but to insist, as has been done, that a switch used for isolating should not be called an isolating switch seems clearly contrary to an established and useful practice.

9.2.1 Accident record

Improper use of isolators and isolating switches over the years has caused a comparatively large number of accidents in high voltage equipment. Some of these are reviewed briefly below.

Breaking load current. Many older engineers have, either in emergency or by accident, broken load current on a plain break isolator or switch, probably operated by a hook stick, and got away with it; but it was always a highly risky operation and with the heavy currents and inductive circuits now common it is suicidal.

Breaking magnetising current. It might be thought, from the definition of an isolator, that it could be used with impunity to break the magnetising current of an off-load transformer, but this can be very dangerous. Some years ago an engineer in a power station inadvertently opened the gang-operated isolators in a steel cubicle. It was at the end of a day when there had been continual trouble with the cooling water and other matters and, being very tired, he had omitted to check that an off-load transformer fed by this circuit had an alternative supply. There was a three-phase flashover and the resulting short-circuit blew the steel door off the cubicle and killed him. There was here a failure of organisation, but the accident might have been prevented by a system of interlocking.

Closing onto faults. There is always a danger when energising a circuit that it will be faulty, and that a short-circuit will immediately occur. This has been the cause of many accidents, of which the most frequent have almost certainly been from closing onto a cable which had been deliberately earthed for repair or testing. With the high fault levels now common it is clear that only a circuit-breaker or switch specially constructed for this purpose should be used.

Access to live equipment. Isolators or non-automatic isolating switches are regularly used to ensure a positive break in a circuit on the live side of apparatus on which work or testing is being carried out; a circuit-breaker alone is not considered suitable for this purpose, as it may itself require isolation for maintenance. When an isolator is used in this way it should be locked (or padlocked) in the open position.

At times, however, the point of danger is in the isolator (or switch) unit itself, as when access to the enclosure seems necessary for testing, particularly on outgoing cables, or for cleaning insulators, or for access to small wiring. This risk must be prevented, but as the isolator is itself the means of ensuring safety and it is impracticable to add isolators to isolators, it is usual to rely on key interlocking (or padlocking) of enclosures, backed up by a suitable routine and supervision.

A particular form of this latter danger, which at one time caused many accidents, was access to the plug sockets of oil circuit-breakers

with vertical or horizontal plug-in isolation. An accessible socket provides an almost irresistible temptation to a conscientious workman to wipe the inside of the tubular insulator. This usually results in a flashover and serious injuries – though rather surprisingly they are usually not fatal. Flashover also occurs from attempts to insert earthing equipment into live sockets and also occasionally from the use of unsuitable live-line detectors.

9.2.2 Constructional precautions

Having briefly reviewed some of the more common hazards with this class of equipment, it is now necessary to consider what precautions should be taken in the construction and selection of equipment to reduce the risks to a minimum.

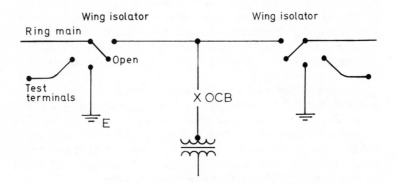

Figure 9.1 Isolators in the high-voltage ring main unit

One of the greatest risks with high voltage equipment is that from closing onto an earthed or faulty circuit and this has caused a continuing series of serious accidents including multiple fatalities. It is clear from this experience that, with the present high fault levels, circuits should be energised through an automatic circuit-breaker or a switch capable of closing on to and carrying the prospective current without danger until it is cleared elsewhere. This does not present much difficulty on large switchboards, but in suburban and rural areas where capital costs are high compared with the expected revenue it was in the past common to control a local transformer by means of a circuit-breaker with 'wing isolators' of virtually zero making capacity (Figure 9.1).

To prevent a live cable being inadvertently earthed, it has been usual to provide a gate or other device to stop an isolator or switch being moved quickly through the 'open' to the earthed position. These devices usually prevent accidental movement beyond the open position until some specific action has been taken, such as rotating a disc or shooting a bolt which cannot be moved until the open position has been reached.

Such devices have, however, frequently proved mechanically unreliable as a result of light construction, poor design or excessive backlash in the moving parts. A contributory cause has been insufficient clearance between the open and earth positions of the blades, allowing them to approach within sparking distance of the earth contacts without the interlock failing completely.

The mechanical integrity of both new and used gear should be investigated carefully. On one occasion the mechanism had become so strained that an engineer could only free it before reclosing by releasing the interlock and in so doing he inadvertently moved the blades towards 'earth' causing a serious explosion in which several persons were injured.

It will be noted from Figure 9.1 that the blades may be moved beyond the earth position to a position where a connection to the cable terminals may be made for testing. It is important that access to the test terminals does not expose conductors on other circuits which may be live, even though they are not easily reached. Accidents have been caused by a strand of copper wire used in tests falling into the switch or being poked in.

Oil circuit-breakers in substations are commonly isolated by plugging into sockets on the fixed enclosed busbar system. These need never be used to open a live circuit, and in fact it is simple and usual to interlock the withdrawal gear so that this cannot be done. There are, however, two particular hazards with this class of equipment. In the first place, the socket openings, or ports, should be automatically covered by shutters when the switch is withdrawn and these should be locked in place, either automatically or with a padlock. It is frequently necessary, however, to obtain access to the outgoing cable sockets for testing and it should be possible to open these without unlocking the cover over the live busbar contacts. This is now standard practice but the interlocking mechanism is not always satisfactory and some older shutters still exist which cannot be operated in this way. Such equipment has often been left on, or relegated to, the fringe of the high voltage system and forgotten.

Circuits connected to ironclad switchgear with plug-in isolation are normally earthed through the circuit-breaker by means of extension plugs and special earthing equipment. This often lies unused for long periods and it is perhaps never necessary to use it at all. When it is

required it is not infrequently found to be very difficult to fit, in the confined space available, and occasionally fitting is quite impossible because of some error of design or assembly. It is important that everyone who may have to use such gear should know how to do so and that it should be tried to ensure that it will work.

An isolator as defined above should never be used to make or break load or magnetising current. Where this cannot be ensured by interlocking, a switch (usually oil-immersed) should be installed, which is defined in BS 2631: 1955[1] as 'a device suitable for making a circuit in oil under normal and abnormal conditions, such as those of a short-circuit, and capable of breaking a circuit in oil under normal conditions'.

The details of the duty of such a switch and the test procedure are set out in the relevant British Standard, but the requirements may be briefly summarised as follows. The switch must be capable of safely breaking any current up to the rated normal current, including transformer magnetised current and cable charging current, and it must be capable of safely making and carrying for a short time any current up to its rated making capacity — at rated voltage. The switch should, however, not be used to break the circuit immediately after making and it should not be closed more than twice on an overload or short-circuit without being examined to see if any maintenance is required. This is no more than common sense.

The specification further recommends that, for switches of rated voltage above 11 kV or a rated making current in excess of 33.4 peak kA, a mechanism should be used in which energy is stored during the first part of the movement of the operating handle and released later in the same operation, so that the speed of the moving contacts is substantially independent of the force exerted by the operator, i.e. that the movement of the contacts cannot be restrained by the operator after the start of the current flow. This is extremely important as many of the worst accidents have occurred when competent men have closed isolators or isolating switches hesitatingly or, realising that they have made a mistake, have stopped halfway, or attempted to pull back. Magnetic forces commonly make this danger more serious with prospective high fault currents but this hazard has little relation to voltage and the writer feels that this feature should always be included irrespective of voltage and preferably irrespective of current. By far the greater number of accidents occur at not more than 11 kV.

It sometimes seems necessary to open enclosures housing live equipment to get at details associated with other apparatus, such as small wiring, or the heads of bolts. There is a great temptation for fitters to do this without authority, but the reason for doing so can be largely eliminated by careful design.

9.2.3 Conclusion

In the previous paragraphs an attempt has been made to indicate some of the constructional requirements which should be borne in mind when selecting or overhauling h.v. equipment, but the operation of high voltage switchgear can never be made entirely free from hazard and design features cannot take account of all possible mistakes. Correct design must, therefore, be supplemented by proper standing instructions and properly planned work, including suitably worded permits to work or to test.

Many of the constructional requirements outlined here can be related to statutory regulations either in relation to construction or methods of work, and with increasing loads, shortages of skilled staff and increasing fault levels the implied requirements must become more far-reaching. It is certain that some existing equipment, though good of its day, may fall below present practice, and while this exists the importance of carefully planning and supervising work is particularly important.

9.3 Medium voltage equipment

The *function* of a piece of equipment may be defined fairly broadly as the job it is intended to do in a particular situation, whereas the *duty* of a switch or fuse must be defined in a more restricted manner in relation to its ability to operate under the stresses imposed in service, and in particular on its ability to open (or close) on live circuits. These aspects will be considered separately, but it will be seen that they are not the only criteria to be considered in deciding on the suitability of a switch or fuse.

9.3.1 Definitions[2,3]

Isolators and isolating switches are primarily used to disconnect circuits and apparatus on which work is to be done — and to keep it disconnected — or to separate two parts of a network, e.g. to prevent overloading, or for operational reasons. As for high voltage equipment, an isolator is defined as a 'mechanical device capable of opening or closing a circuit under conditions of no load or negligible current;' the distinction between load and current seems rather obscure.

In practice, isolators as distinct from isolating switches are little used today on medium voltage circuits – probably the principal exception has been the use of plain-break knife switches for isolating the circuit-breakers on open-type switchboards. These should properly be called isolators but the term knife switch is so well established that it would be pedantic not to use it (see section 9.2).

Switches have been defined as mechanical devices for making and breaking non-automatically a circuit carrying a load and, very important, they should have a declared making and breaking capacity. These ratings are no more than normal overloads and it is clear that currents in excess of these values should be prevented, e.g. by a cartridge fuse or by selecting a switch with a higher normal current rating. The rating of the switch must therefore be related to the protection of the circuit as well as to the load current of the apparatus controlled. In particular a switch may have to make, and carry for a short time, the initial current rush of filament lamps and induction motors. Instantaneously this may be several times the continuous current rating, but it must not blow the fuses. Strictly, therefore, the making capacity and fusing should be related to the initial current. Provision for this is usually made in heavy duty combined units[4] but not in ordinary everyday units. This will seldom introduce any great difficulty, however, except close to medium-sized or large transformers where the prospective short-circuit currents may be high, since the making capacity is, in fact, usually sufficient. The difficulty could probably be met without substantial redesign or increase in cost, by specifying a making current several times greater than the breaking current to take account of the current inrush. A change in the requirements relating to tests would be necessary to give effect to this.

There are two reasons why simple isolators should rarely be used. In the first place, practically any isolator may be operated on load inadvertently, even by skilled persons, unless it is interlocked with a switch, circuit-breaker or contactor; and secondly, where isolators might be used to disconnect a starter or controller, experience shows that it is most important for the isolator to be opened whenever the machine or process is stopped. A machinist or plant attendant cannot be relied on to do this if he normally shuts down by operating the controller. It is therefore better that he should shut down by operating an *isolating switch* as he may, and at times undoubtedly will, do this while current is still flowing.

At one time links of construction similar to fuse-holders in locked compartments were used for isolating controllers, on the assumption that they would only be used by electricians, but this practice has been discontinued with the realisation that it is important to isolate when a machine is stopped to prevent mechanical accidents from false starts

caused by electrical faults in the controller or by a person inadvertently knocking a controller handle or push-button, or even operating the wrong controller.

9.3.2 Location[5]

Means of isolation must be suitably situated 'geographically' and also be in the proper place in the electrical network. Regulations require that, where necessary to prevent danger, efficient means suitably located shall be provided for cutting off all pressure from every part of a system, and that for every motor, converter and transformer, the pressure shall be cut off from all apparatus used in connection with it, e.g. starters, controllers and instruments, and adequate means shall be taken to prevent any conductor or apparatus from being made live

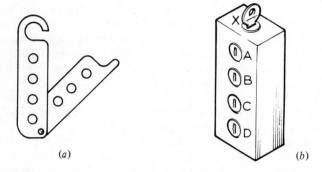

(a)   (b)

Figure 9.2(a) Device for securing a switch or isolator by means of several padlocks. (b) A key block used to control a switch or isolator by several keys. When, and only when, the appropriate keys are inserted, turned, and thereby trapped in each of the locks A, B, C and D, the key in X is released and can be inserted in a lock integral with the switch, which when operated releases the handle and allows the switch to be closed. This key is trapped so long as the switch is closed, and the keys of A, B, C and D are therefore also retained

while anyone is working on it. Obviously, similar considerations apply when anyone is working on a machine that is electrically driven. In virtually all situations this isolation is best provided by an isolating switch, not a circuit-breaker or contactor, nor an isolator, unless this can be interlocked with an associated switch, circuit-breaker, or contactor so that it cannot be used to interrupt current.

The isolating switch should be placed conveniently for the person who has to use it, and the isolating switch for a controller should be as close as possible to the controller, so that the machinist or attendant has no excuse for not opening it when shutting down; this also makes it less likely that anyone will inadvertently close (or open) the wrong switch.

An isolating switch close to the controller is usually the best way to protect an electrician, machinist or maintenance engineer working on electrical or mechanical equipment. Occasionally, however, on large remotely controlled plant, additional points of isolation are required; the rule should be that either the isolating switch is close at hand or it can be locked open, the key being retained by the person at risk or who issues a written permit to work.

Figure 9.2(*a*) illustrates a device enabling several people to use their own padlocks. Where figure key locks are used (Figure 9.2(*b*)) a multiple key block is provided. If isolating switches are used to disconnect crane trolley wires and similar conductors it is desirable not only to supply locking facilities, but also to use switches which short-circuit and earth the exposed conductors besides isolating them. The point of isolation should in no case be unduly remote from the work to be done or otherwise when making a small adjustment, e.g. to an overload setting, men will take a chance and unnecessarily work on or near live conductors, particularly if a series of adjustments and trials may be necessary. In the case of trolley wires, it may be necessary to work some distance away from the switch, but it should, if possible, be in view and should *not* be in a locked switch house or substation.

The proper position in the network or circuit for an isolator or switch depends on its function or purpose. It should normally immediately precede a controller, and where several motors are needed to operate a single machine, each should have its own isolation unless they are all controlled simultaneously by the same controller. Seperate isolation is essential if some of the motors can be used without the others, and is very desirable for allowing an extensive system to be split up for testing section by section. An example of this arises where medium or high voltage contactors are operated by lower voltage control circuits; it should be possible to test the control circuits with the high voltage circuits isolated and locked off. Failure to provide adequate segregation and isolation of circuits at main distribution points may gravely interfere with maintenance work.

Another problem arises from sneak circuits whereby live connections may be found in control apparatus, the main supply to which has been disconnected. Three examples may be quoted. In the sequence control of a conveyor system it is necessary to start the last conveyor first and work back to the first, to ensure that there will be no pile-up at an

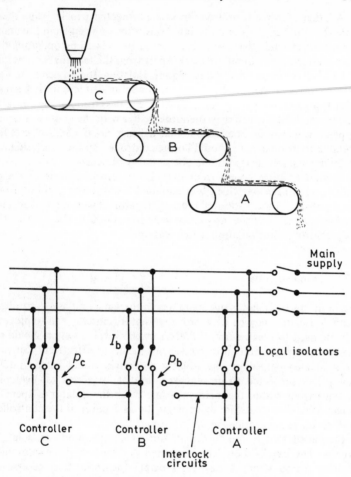

Figure 9.3 Conveyors in sequence. Unless special provision is made for isolation, pilot terminals P_b may be live while the local isolator I_b is open, and similarly at C

intermediate point (Figure 9.3). To ensure this, each controller includes a pilot circuit from the one below, which must either pass through auxiliary contacts on the main isolating switch or through an auxiliary switch. The latter precaution is more convenient for testing, but if it is adopted a clear indication of the fact is essential. The difficulty may be overcome by an interlock which can be defeated by an authorised person, but must be reset before the equipment is returned to service.

Another example is provided by crane protective panels. When these were first introduced a main isolator was often omitted from the crane cab on the grounds that the circuits could be isolated by operating the main contactor (or circuit-breaker). Apart from the fact that contactors and circuit-breakers themselves require isolation for servicing, it had been overlooked that pilot circuits for the limit switches must be supplied from the live side of the contactor and pass through the contacts on the controller, which may therefore be live with the contactor open. In passing, it may be pointed out that lighting circuits and sockets for portable tools on a crane should not be controlled by the main isolator, otherwise maintenance work may be seriously impeded.

A third example comes from the 110 V control cubicle of a high frequency furnace on which was mounted a wattmeter supplied from the input side of the associated motor generator. It was overlooked that the back contacts of the instrument were live at 400 V when the 110 V main control circuit isolating switch was open.

9.3.3 Combined units

Where means of isolation are combined in the same unit as a controller, a simple isolator may be used provided that it is interlocked to prevent it being used to break current. Preferably, however, a switch should be used because this is the most convenient means of effectively shutting down in an emergency. The isolator or switch should be separately enclosed or protected to prevent exposure of, or access to, live connections from within the main enclosure when the isolator is open. A danger otherwise exists; for example, if the back cover of the controller is off to get at connections in the controller.

Combined switch and fuse units are commonly used for isolating circuits. The switch should be protected as discussed above, although there is not so strong a case for separate enclosure. It is necessary, however, to consider the relation between the rating of the switch, the fuses and the prospective current. For ordinary combined units the making and breaking capacity tests are on the same basis as those quoted earlier in this article, so that, unless it can be established that the cut-off current of the fuses is not greater than these values, higher rated switches must be selected or, preferably, heavy duty composite units should be used.

Isolators or switches associated with the controllers of large motors and some other equipment may well be closed on what is very nearly a short-circuit if the automatic device for bringing the starter to neutral should fail, and in some other conditions; also, should a motor stall and the overload protection fail, the attendant is likely to attempt to clear

the circuit with the isolator. For this reason heavy duty combined units of the type described are recommended. Alternatively a simple switch may be interlocked with a contactor or circuit-breaker, integral with the controller, so that it cannot be used to break current. If the contactor or circuit-breaker welds up the duty of clearing the circuit is thrown further back. Failure to take account of these possibilities has caused considerable trouble, including serious accidents at some steel works.

9.3.4 Construction of units

It is not intended to review the whole subject of switch design, but to refer to a number of details which are frequently overlooked. Clearly the mechanical parts must have adequate strength and resistance to shock and fatigue, and the quality and thickness of insulation and creepages and clearances must be adequate.

High breaking capacity cartridge fuses should always be used except for lightly loaded circuits where the prospective short-circuit current is small. Even if a larger fuse is substituted for the appropriate size, this should not endanger the switch and fuse unit, even if the *circuit* is no longer adequately protected, since the maximum size is determined by the fuse holder and contact arrangements.

Facilities for cabling should be examined, bearing in mind that at times over-sized cables must be used and must not be damaged while being inserted, and preferably it should be possible to carry out routine inspection and testing without removing barriers and arc chutes which may not be put back. It is also as well to remember that work on live gear is occasionally unavoidable, and a design which facilitates fitting temporary protection over live parts is valuable.

Covers are usually interlocked, but it is desirable that skilled men should be able to defeat interlocks for the purpose of testing. If this facility is not provided some vital feature of the equipment is likely to be removed or forced so that access may be obtained.

Where a number of units (switches, fuses or isolators) have been connected to each other, or to busbars, many accidents have been caused by small articles such as pieces of wire, screwdrivers, penknives, nuts and washers falling from one unit into another — or being poked through apertures — and thus causing flashover. All apertures between compartments should be effectively sealed.

It is important that a switch which appears to be open shall not be partly or fully closed. This has caused a great deal of trouble and many older switches have excessive backlash, so that after a little wear it is possible for the contacts to remain closed with the handle poised in an

intermediate position so close to the fully open position that its true position can be mistaken. Such matters have been covered in various British Standards.

9.4 Conventional control systems

9.4.1 Classification of risks

It is possible to classify most of the risks as follows:

(a) Inadvertent starting of machinery which may endanger the life or limbs of fitters, millwrights, tool setters, machine operators and others who are working on the machines.

(b) Failure of machines to stop when they should do so, due for example to inefficient brakes or lack of suitable over-running gear.

(c) Excessive speed and over-driving which may, for example, cause the bursting of grinding wheels or mechanical damage to persons and machines.

(d) Unauthorised access to danger areas such as magnesium grinding and electric cable-testing cubicles, or X-ray departments.

(e) Failure or inadequacy of electrical safety devices.

The above list gives, on the whole, an adequate picture of the likely risks. However, it is not exhaustive and special hazards may arise from time to time which cannot be conveniently classified in this manner.

9.4.2 General principles

Failure to safety has been mentioned above. The guiding principle is clear though the method of application will vary according to circumstances.

9.4.2.1 Excessive backlash and false position

This matter, which was discussed in the previous section in relation to isolating switches is important to the control of machines; for example when a machine had unexpectedly started and had killed a man it was uncertain whether the isolating switch had been properly opened as it was quite easy to bring the handle very close to the 'off' position without the contacts separating. The backlash should therefore be limited to a comparatively small movement, so that the switch contacts must separate long before the open position is reached.

False position also arises in the setting of limit switches and the return motion of automatic machines, as a result of which safety devices on return motions may fail to become effective at the proper time.

9.4.2.2 Standardisation and position of controls

It is very important for controls to be standardised within any one works, e.g. the arrangement and direction of rotation of the barrel controllers in a crane cab, so that mistakes do not arise when a worker is transferred from one machine to another.

Attention may be drawn here to the inconvenient and sometimes dangerous positions in which built-in controllers and push-buttons are fitted in some machine tools. Little attention appears to be given to the requirement that they must be easily worked by the person in charge of the motor, i.e. the operator.

Electricity Regulation No. 12 requires that 'every electrical motor shall be controlled by an efficient switch or switches for starting and stopping, so placed as to be easily worked by the person in charge of the motor.' There is a somewhat similar requirement in the Woodworking Machinery Regulations.

9.4.2.3 Motor starting torques

Where a motor has to start against a load it is essential that it should have ample reserve of power and adequate starting torque, otherwise if left switched on in the starting position it may remain stalled, and then start suddenly if vibration or any other cause reduces the mechanical resistance to starting. Thus, for example, an induction motor provided with a rotor starter should have adequate torque on the first working notch.

9.4.2.4 Overload and low-volt releases

The probability of accidents arising from stalled motors is increased if there is unauthorised tampering or failure to maintain overload devices. Experience shows that the type of overload release which short-circuits the no-volt release is apt to be inoperative owing to dirty contacts. On the principle of 'failing to safety', devices should always operate by opening a circuit. On one occasion it was found that only two or

three d.c. overload releases in a large engineering shop worked satis-factorily.

With dangerous machines it is also necessary to have efficient low-volt releases, so that they do not restart unexpectedly when the power supply is resumed after a failure which has occurred either in the public mains or through fuses or switchgear operating in the works installation.

Faceplate starters in particular are liable to stick in the 'on' position owing to dirty contacts or weakened springs, and sometimes they have no 'off' position, the first stud being a starting stud. All things con-sidered, a series contactor between the isolator and starter with an independently reset pushbutton is often the best choice for overload and low-volt protection, and is not necessarily very expensive. The advantage of this arrangement when combined with emergency stop and reset buttons is mentioned below.

The fact that starters are sometimes found with the handles tied up to prevent them tripping is usually an indication that the motor, the starter, or the setting, or all three are unsuitable.

Should the supply fail, all isolating switches should be immediately opened, since there is always a danger that where there are a number of starters at least one low-volt device is ineffective.

9.4.2.5 'Inching'

When preparing or setting up some types of machinery, a slow-running or an 'inching' device is necessary. Attempts to inch with an ordinary barrel controller by quickly making and breaking the first notch, not only leads to excessive wear and maintenance costs, but may cause accidents. For example, a man had to feed strands of wire through the slowly revolving head of a wire rope-making machine with one hand while controlling the speed of the machine with the other. Although he had done this on many occasions, he inadvertently caused the machine to run a little too fast and lost contact with the control handle, and was drawn into the machine. Fortunately it was running very slowly, and he was released before he suffered very serious injury. The provision of a properly located slow-running controller of the 'dead man's handle' type would have prevented this accident.

An alternative to a slow-running button is a true inching button designed so that the machine moves forward a short distance and then stops after each depression of the button. This may be safe in circum-stances where a slow-running button would be dangerous, but such inching buttons should be in a position convenient for the persons actually carrying out the work, and should be so placed and constructed that they cannot be confused with the normal start button. Occasions

have arisen where the inching control was so linked with the main supply that both controllers could inadvertently be applied at once, thus causing incorrect operation of the machine; this, of course, should be avoided, and, in fact, a throw-over switch which will prevent the main starter being used when the machine is being controlled by inching is advantageous.

9.4.2.6 Stability

Equipment may be left in an unstable position, so that a slight jar or other inadvertent movement may cause the machine to start. This is largely covered by Factory Electricity Regulation 3(*d*), which states that every switch, etc. shall be so constructed and arranged that it cannot accidentally fall or move into contact when left out of contact.

The defects, however, may not be in the electrical equipment but in some part of the mechanical equipment, such as gears, clutches, brakes or fast and loose pulleys.

9.4.2.7 Starting procedure and grouping of controls

Where large machines or groups of machines are necessarily controlled from a central point so as to ensure co-ordinated operation, it may not be possible for anyone at the control station to see all the dangerous positions in which another person may be. Some system must then be adopted for ensuring that everything is clear before the machinery is started.

Where there is a sufficient number of operatives this can be effected by the use of lock-in pushbuttons of the type described below. In other cases it may be necessary to rely on warning or other signals to which, where possible, there should be an agreed reply. This applies particularly to large printing machines and to long conveyor systems, some of which may extend for several hundred yards. The unnecessary grouping of controls is, however, undesirable and may lead to mistakes, particularly when the operations of the machines or processes controlled are not closely associated and may in some circumstances be contrary to statutory regulstions. This does not, however, prevent the grouping of contactor panels with local pushbutton control or master controllers.

It is also opportune to mention the rather obvious point that where a man is driving a travelling machine such as a furnace charger or travelling floor crane, he should be able to see where it is going, and the control station should not be tucked away in a corner.

9.4.2.8 Pushbuttons

Contactor control with pushbutton operation has many advantages, but there are also disadvantages. It has been mentioned above that an inching button should be so constructed and placed that the main start button cannot inadvertently be used instead. The same also applies to the stop button. Very frequently stop and start buttons are similarly constructed and placed side by side, and on occasions they are not even clearly marked. Generally speaking the stop button should be so constructed that it cannot very well be missed, even if one hits out at it rather hurriedly; while, on the other hand, a start button should be recessed so that it cannot easily be depressed except deliberately (see Figure 9.4).

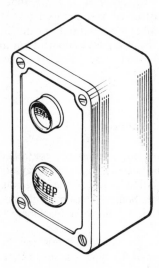

Figure 9.4 Mushroom-headed stop button. The start button is shrouded; this prevents it being pressed accidentally

The pushbuttons themselves also suffer from certain maladies not common to other forms of control. Plastic buttons may swell in contact with oil or other liquids, and should a start button stick it may result in a machine stopping when the stop button is depressed and restarting immediately the finger is removed. There is also a tendency for cutting fluid to trickle down a man's hand, along his finger and on to a pushbutton, and it may thus enter the mechanism and cause electrical breakdown. For example, internal grinders have inadvertently started on several occasions owing to earth faults which are believed to have occurred in this manner. There are various precautions which can be taken; for example, the pushbuttons may be so arranged that they point slightly downwards, causing any liquid to run off rather than run

in, or a bridge incorporating a dummy buttons may be placed across the front of them, so that the finger pushes a dummy button which, in turn, pushes the real one. Both these precautions have been found effective.

Further points relating to pushbutton control are dealt with in the following paragraphs.

9.4.2.9 Emergency stops

Substantially constructed pushbuttons are frequently used to provide for the stopping of machinery in emergencies, but the manner in which these pushbuttons are connected to the controllers is often undesirable. For example, it is not unusual for the button to open-circuit the low-volt release coil. This is particularly dangerous where the driven machinery and the stop button are at some distance and perhaps out of view of the controller, whether automatic or hand-operated. Should the machine be stopped by the pushbutton, either in emergency or for the purpose of making adjustments, it is then possible for another person to start it immediately the button is released. In addition to this the motor can generally be run up to speed in either the first or second position of the starter, and will continue to run so long as the handle is held. Both these have caused accidents. This method of connection should be used only with a starter of the 'no-volt, no-close' type in which it is impossible to complete the main circuit with the low-volt coil open-circuited.

(a) Emergency stops on extensive machinery

There are considerable advantages in using an auxiliary series contactor operated from the emergency stop buttons only, and these stop buttons should be of the locked or stop-and-reset type. Thus, if there are several such buttons and one is depressed, the circuit must be reset at that point and no other. This arrangement is very common on newspaper-printing machinery where a number of men are attending a single machine and any one of them may stop it to make some adjustment should, for example, the paper tear or curl. Since each man has his own stop and reset buttons no person can restart a machine which someone else has stopped. This, of course, implies good team work as well as correct selection of equipment.

(b) Lock-in buttons

There are generally two types of lock-in button: in one it is necessary to depress the button and turn it if one wishes the machine to stay

definitely stopped; in the other stop and reset buttons rest on opposite ends of a lever so that when the stopbutton is depressed the reset button is definitely knocked out and vice versa. Both prevent the risk of a machine restarting when pressure on the stop button is released, but the first form has the disadvantage that in an emergency or moment of excitement a person may forget to twist as well as push the button.

(c) Trip wires

It is not always easy to cover an area of risk adequately by the provision of pushbuttons. This difficulty may sometimes be overcome, as for

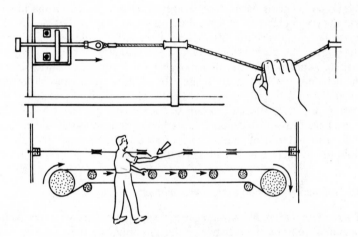

Figure 9.5(a) Installing a trip wire. The diagram shows the wire installed correctly so that it can be operated by a pull in either direction

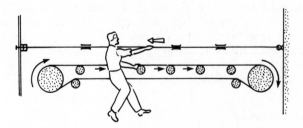

Figure 9.5(b) Installing a trip wire. Incorrect installation, the wire is inoperative if pulled in the direction shown

example on a conveyor, by suspending a trip wire along the line of the plant; pulling this at any point will bring the machine to rest. It is important, however, that arrangements should be made so that the machine is not restarted without ascertaining that all is clear at the point at which the trip wire was pulled. For this reason it may be necessary to sectionalise the protection and use a number of comparatively short trip wires each acting on a separate lock-out tripping device. The correct method of installing a trip wire is shown in Figure 9.5.

9.5 Limit switches

Limit switches are used for four main purposes[6].
1. To prevent moving parts of machines from over-running.
   Typical examples are the hoist limit switches of cranes and the ultimate switches at the top and bottom limits of the travel of lifts.
2. To ensure that moving parts are brought to rest at the correct place, as by the levelling switches on automatic lifts. Limit switches may also be used to stop or reverse the motion at the end of a stroke or to control other functions on machine tools and such plant as injection moulding machines.
3. In association with mechanical and electrical interlocks, to allow certain operations to be performed only when parts of a machine are in a particular position and to prevent them from being performed at all other times. Examples are limit switches on lift gates and on guards on milling machines, failure of which might lead to immediate danger.
4. To stop machines in certain predetermined circumstances, as when a guard is opened or when the material on a conveyor or a sequence-controlled machine is out of place or missing. The failure of a limit switch on a guard is almost always dangerous, and there are often special hazards in other places where they are used for this purpose.

It will be seen from these examples that there is a variety of conditions which limit switches may be required to fulfil, and it is not to be expected that any one type of switch will be suitable in all circumstances. A general distinction may be made between those switches which are used to control a production sequence, where failure may not necessarily cause danger, and those on which persons habitually rely for their safety.

Any safety switch must be of robust construction and stand up to the conditions of use including foreseeable misuse. It must be possible

to set it accurately and the setting must not be disturbed by maintenance work or by wear and tear so far as this is practicable. In particular it must not fail either by the breakage or weakening of a spring. To prevent this, where springs are used to obtain a quick break, there must be positive mechanical back up action to ensure that the switch operates within the safe limits.

In many circumstances an over-run could damage the switch or mechanism and steps must be taken to prevent an over-run or render it innocuous.

Further examples and accounts of some accidents will be found in Reference 7.

When a limit switch is first installed, replaced or adjusted, this should be done with the accelerating springs removed to make sure that the positive drive will become effective should the spring fail without delaying the safety action. Secondly, when once a limit switch has been fixed in position and correctly adjusted, it should be secured by well-fitted dowel pins, so that if removed it will go back in exactly the same position. One may hope that limit switches will eventually be made with such accuracy that switches of the same make can be interchanged without upsetting the adjustments.

9.6 Micro-switches

The terms micro-switch and micro-gap switch are often used in a misleading way and are in danger of losing precise meaning.

The micro-gap switch stems from the discovery that a very small gap or break between suitable contacts is adequate to break alternating currents up to a few amperes in non-inductive circuits, such as for lighting or resistance heating. The matter was investigated by Professor W.M. Thornton[8]. For thermostats controlling water heaters, the gap is limited to 0.005 in. with contacts $\frac{1}{4}$ in. to $\frac{3}{8}$ in. diameter. With this small separation, ionisation of the air in the gap does not persist. The switches used normally have a snap action, and the gap is limited by a back stop. The small contact separation is essential to satisfactory operation and if it is larger, arcing in the gap may cause sufficient ionisation for the current to re-strike.

It does not necessarily follow that such a switch is suitable for general use in safety circuits. The spring may fracture, the inductance of operating coils may prevent a clean break, and it may not be possible to ensure exact adjustment of the operating mechanism so that the switch mechanism is not strained; overtravel is particularly damaging. A further weakness of switches with such small contact separation is that the gap may break down if there is a high voltage transient, such as

may be caused by current chopping in a parallel circuit. Russell Taylor[9] measured transients as high as 11 000 V d.c. and 3000 V a.c. which were caused in this way in relay-coil and contactor-coil circuits. It is clear, therefore, that switches of this type should be used only where it is known that such dangers will not arise.

The advantages of the micro-gap switch need not, however, be disregarded. The essential point is that small alternating currents in circuits of low inductance may be broken by very small contact separation, and this has many useful applications but they are unlikely to be satisfactory for industrial safety applications.

'Micro-switch' is a term now often applied rather indiscriminately to any small switch in a sealed box or capsule; in fact it is sometimes used for limit switches, which may be large or small and which have entirely different characteristics. Because properly designed small switches with positive action of one sort or another are frequently and correctly used for interlocking and similar purposes in safety circuits, this loose use of technical terms is very regrettable and may lead, directly or indirectly, to danger, since instructions and purchasing orders may be misunderstood. The danger of using a spring-operated switch without mechanical back up must, however, be remembered and when the contact separation is very small a slow break switch with wider separation as final break is desirable.

9.7 Mounting and operation of limit switches

It is important that safety switches should be opened positively so that they cannot weld or stick, or be 'faked'. This is referred to below under 'guards'. As already stated, it is also important to ensure that an overrun will not damage the switch.

In Figure 9.6(*a*), the switch is opened positively and fails to safety if the spring breaks, whereas in Figure 9.6(*b*) with the operation reversed it is not positively opened and does not fail to safety. In addition, overtravel on closing will damage the switch, the lift of the cam being critical.

Figure 9.7 shows a switch with a central driving spindle which may have a return spring, although this is not always necessary. The action is positive and the switch will not be damaged by substantial overtravel. If it is operated by a cam and roller, or by a ramp, it can easily be made 'non-self resetting' by omitting the spring.

In Figure 9.8(*a*) the cam forces the switch open as the guard is opened. In Figure 9.8(*b*), however, the action is not positive and Figures 9.8 c and d shows how the interlock may be defeated by sticking or being 'scotched'.

In other circumstances damage to a switch can be prevented by the design and mounting of an operating cam as in Figure 9.9a. The safe limits of use are indicated in Figures 9.9b and 9.10. A and E are the limits which must not be exceeded if the switch is not to be damaged,

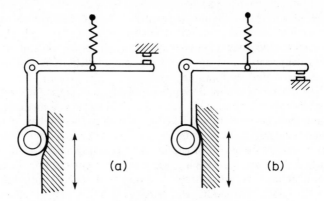

(a) (b)

Figure 9.6 Operation of limit switch

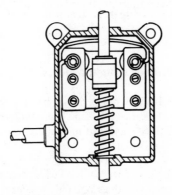

Figure 9.7 Limit switch with free overtravel

while B and D must be over stepped if the switch is to function correctly. Figure 9.10 shows the same limits for a different design of switch.

9.8 Faults in conventional controls

In this section we are primarily concerned with failures or deficiencies which may cause a machine or process to operate incorrectly and thereby dangerously. Simple examples are a fire alarm or an emergency

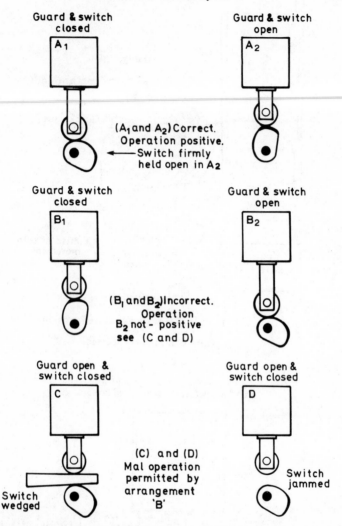

Guard & switch closed

A₁

Guard & switch open

A₂

(A₁ and A₂) Correct.
Operation positive.
Switch firmly
held open in A₂

Guard & switch closed

B₁

Guard & switch open

B₂

(B₁ and B₂) Incorrect.
Operation
B₂ not - positive
see (C and D)

Guard open &
switch closed

C

Guard open &
switch closed

D

(C) and (D)
Mal operation
permitted by
arrangement
'B'

Switch
wedged

Switch
jammed

Figure 9.8 Correct and incorrect use of limit switch

stop button. Circuits which are normally closed and operated by opening a switch will operate on a failure of supply or broken circuit and to that extent fail to safety; but the switch can be defeated, accidentally or intentionally, by a short circuit. When the switch is normally open it will fail if there is an open circuit. In either case consideration must be given to the security of the supply. An alarm will fail if the supply fails and this is commonly overcome by using direct current

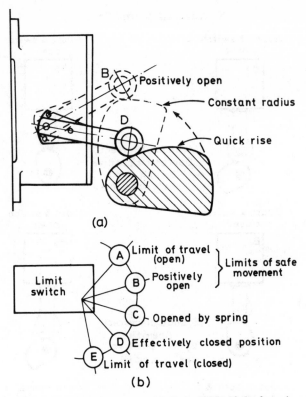

(a)

B · Positively open

— Constant radius

— Quick rise

D

Limit switch

(A) Limit of travel (open)

(B) Positively open

Limits of safe movement

(C) Opened by spring

(D) Effectively closed position

(E) Limit of travel (closed)

(b)

Figure 9.9 Cam operation with limited lift. If the lever is forced beyond A, the switch will be damaged

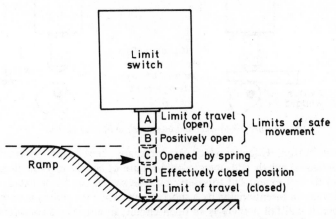

Limit switch

A Limit of travel (open)

B Positively open

Limits of safe movement

C Opened by spring

D Effectively closed position

E Limit of travel (closed)

Ramp

Figure 9.10 Alternative operation of limit switch (switch would be damaged if forced above A)

from a battery supplied by a trickle charger. It does not usually matter, however, if the supply to a stop button fails provided that this also de-energises the motor, but it may be necessary to ensure that a brake is applied automatically, and that a low voltage release prevents the motor from re-starting automatically.

It is clear, therefore, that any control or interlock must be carefully considered in relation to its purpose and the circumstances of its use, and all possible causes of failure, singly or combined, must be considered, including foreseeable abuse.

It is not possible to deal here with all possible occurrences or the steps that have been taken to prevent them. What is necessary is to develop a critical attitude of mind and to take nothing for granted. It is always possible to learn from a skilled craftsman, particularly an erector or maintenance fitter. They know the little things which have gone wrong which the designer never hears about and have been corrected on the spot, the weaknesses of interlocks and those which must be defeated for maintenance to be carried out. Once a fault has been correctly diagnosed it is usually easy to rectify; but one must beware of 'folk lore' in these matters.

The detection of actual or potential sneak circuits is of major importance in safety engineering.

9.8.1 Sneak circuits

In Figure 9.11(*a*) short circuit at B or C will de-energise the operating coil, but a short circuit at A will energise it with the switch open. In Figure 9.11(*b*) an earth fault at A or B will prevent the coil from being energised, while an earth fault at C will have no effect. In Figure 9.11(*c*) an earth fault at A will prevent the coil being energised, while at B it will energise the coil with the switch open.

For most purposes it is best to separate the 'go' and the 'return' conductors so far as is reasonably practicable, and on earthed systems switches should be on the live rather than the earthed side. In Figure 9.11(*d*) earth faults at A and B will together energise the coil with the switch open even though the system is not earthed, while faults at B and C will prevent the coil being energised.

In Figures 9.12 and 9.13 similar considerations are applied to three-phase and single-phase systems with the centre point earthed. In these circumstances the coils will be energised at a reduced voltage which may not be sufficient to actuate them but may be sufficient to hold a switch or contactor closed, or leave the load on an electric magnet precariously supported. The problem may be met by using contactors which release for a small reduction of voltage, if this does not introduce

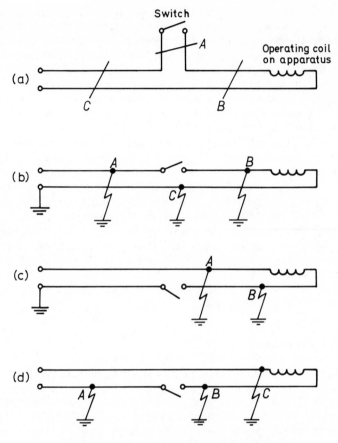

Figure 9.11 Examples of typical 'sneak' circuits

other and greater hazards, or monitoring the voltage with a low volt release relay.

[The above are all simple examples of sneak circuits. In the illustrations the dangers are obvious, but in actual control systems they may be one obscure feature of a complex layout and be very hard to trace. Even more difficult to trace are sneak circuits which are momentarily set up in complex controls arising from an oversight in design and/or the operation of push buttons or relays out of sequence, or in similar ways.]

Figure 9.14(*a*) shows a simple retaining circuit. When the start button D is pressed current passes through the contactor coil and auxiliary contacts B complete the main circuit so that D can be released.

Figure 9.14(*b*) shows the essential details of a controller in which an attempt was made to combine this feature with intermittent operation (pseudo-inching). When D is pressed B closes and is retained as before. When E is depressed the contactor closes but B is not retained because the upper circuit is broken at E. When E is released B should be open before the upper circuit is restored at E and the coil should be de-energised. After some use, however, E re-closed in the upper position before B was released, and an accident was caused.

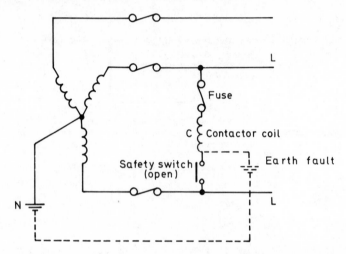

Figure 9.12 Line to Neutral fault with control circuit across phases. (Contactor C energised at Phase to Neutral voltage)

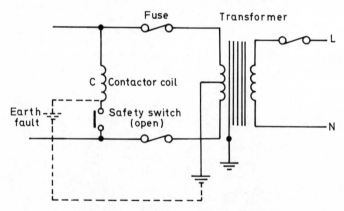

Figure 9.13 Earth fault on single-phase control circuit with centre point earthed C is energised at half-voltage

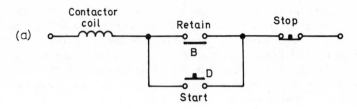

(a)

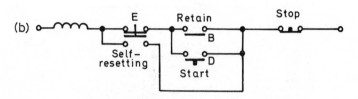

(b)

Figure 9.14 (a) Simple retaining circuit
(b) The circuit combined for pseudo-inching. This arrangement is dangerous as E could close in the upper position before B is released

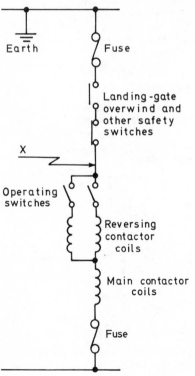

Figure 9.15 Skeleton lift control circuits. With this arrangement if an earth fault is at X it is possible to bypass all safety switches

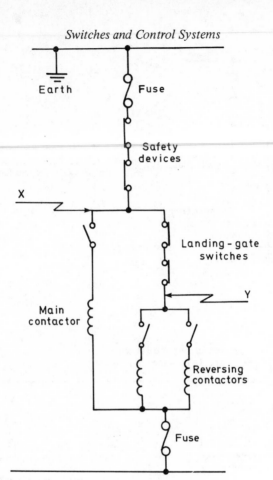

Figure 9.16 Alternative lift control circuit. With the earth fault at Y the reversing contactors could be energised with the gates open

Figures 9.15 and 9.16 show typical skeleton control circuits for lifts. These are normally shunt-connected across the outers of a 3-wire d.c. system or between two lines of a 3-phase a.c. system. Occasionally they are from line to mid-point or line to neutral, but this is not so usual. If one pole of a d.c. system is earthed or, alternatively, unearthed but has an earth fault and there is a further fault, all or part of the safety devices may be by-passed, depending on just where the fault occurs.

In Figure 9.15 it is possible to by-pass all the safety switches if the fault is at X; with the arrangement shown in Figure 9.16, current cannot pass through the reversing contactors with the landing-gate

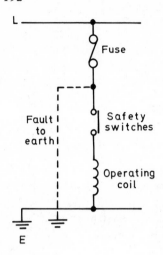

Figure 9.17 Solution to problems in Figures 9.15 and 9.16. The fuse 'blows' and the coil is inoperative

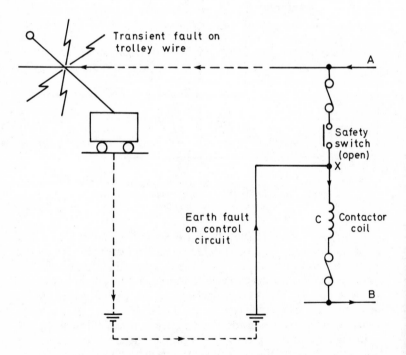

Figure 9.18 Combined faults on insulated system (Contactor C energised)

switch open. If, however, the fault is at Y it would appear that the reversing contactors can be energised with the landing gates open while the main contactor receives its supply in the normal way.

These dangers are eliminated by having the operating coil on the earthed side of the switches as in Figure 9.17.

On insulated control systems, the trouble may be rather obscure because defective insulation, for example at X in Figure 9.18, may not initially give any observable trouble, but unexplained movements of the lift or other machinery may occur when the line A is momentarily earthed and the fault is cleared by a circuit-breaker. This second, transient fault may be in a remote part of the works and may not even be remembered if it was the result of some temporary trouble such as a crane collector arm coming off the conductor. This is particularly likely to happen on unearthed d.c. systems.

These difficulties may be overcome by using a two-wire supply (a.c. or d.c.) with one pole earthed and all operating coils connected to the earthed pole. With this arrangement it is not possible for an earth fault to energise a solenoid or winding with the interlocking or operating contacts open. The use of a safety isolating transformer is an additional advantage and this may be essential where the supply is from a three-phase system with no neutral conductor available, or in many cases where rectifiers are used. Where there are several control circuits, there may be an advantage in using separate secondary windings for different circuits to prevent accidental cross connections (Figure 9.19). Further protection may be given by using earthed-metal-sheated single-core cable and an earth leakage relay (Figure 9.20).

9.8.2 Circuit segregation and safety isolating transformers

A large number of accidents have occurred on lifts and hoists because of the unsuspected existence of sneak circuits and many years ago I recommended that supplies for interlock circuits should be taken, not from the main supply but via a safety isolating transformer so that at least two faults would be required to cause trouble (this is now covered by BS 3535). If the secondary is earthed through an earth leakage relay this can be used to give warning (or initiate appropriate action) on the first fault. It may not be desirable to cut off the supply automatically if this would, for example, leave a lift cage of people stranded between floors. The principle has an application to other control circuits.

For complex control systems separate transformer windings may be used for different circuits, thus further reducing the possibility of sneak circuits (Figure 9.19). In general there should be the minimum possible cross-connections between different control functions as it is difficult

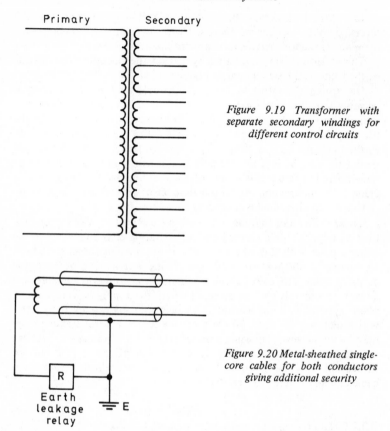

Figure 9.19 Transformer with separate secondary windings for different control circuits

Figure 9.20 Metal-sheathed single-core cables for both conductors giving additional security

Earth leakage relay

E

to foresee the permutations and combinations of faults which may occur and lead to trouble. Single metal sheathed cables for outward and return conductors, with earth leakage protection, gives enhanced security (Figure 9.20).

9.8.3 Rectified supplies

Figure 9.21 shows a so-called multiplier, or free wheeling circuit. During one half-cycle current flows from A through E and B to C but is prevented from flowing through D. On the next half-cycle current cannot flow from C to A, being blocked by the rectifier E. The self induction of B, however, maintains the current which is short circuited through D and continues in the same direction, and the fluctuation is

not great if X is high and the total circuit resistance low. A little consideration will show that similar results may arise with bridge circuits such as Figures 9.22 and 9.23 and this may prevent contactors opening and brakes being applied (or released according to whether they are spring closed and held off by a solenoid or vice-versa) which is illustrated

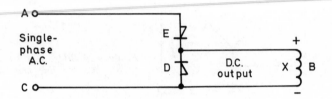

Figure 9.21 Multiplier (or free-wheeling) circuit

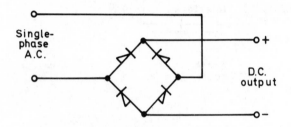

Figure 9.22 Single-phase bridge rectifier circuit

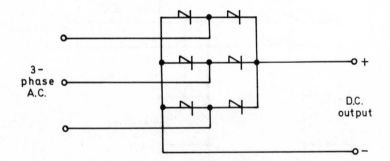

Figure 9.23 Three-phase bridge

in Figure 9.24 in which a bridge rectifier is connected between line and earth. If the right-hand fuse should blow on the occurrence of an earth fault as shown, then the coil will be energised by what is in effect a multiplier circuit and may hold on with the interlocks open.

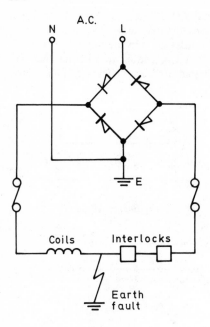

Figure 9.24 Bridge rectifier connected between line and earth

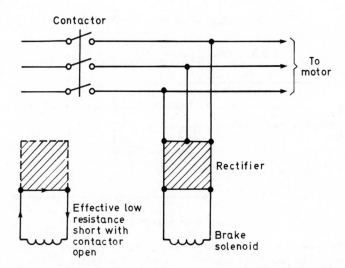

Figure 9.25 Rectified supply acting as a short circuit across a brake solenoid

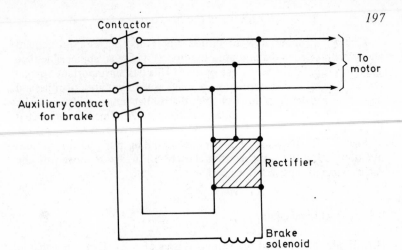

Figure 9.26 Circuit which eliminates the short-circuit produced by Figure 9.26

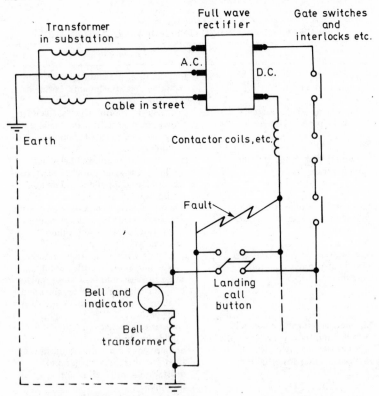

Figure 9.27 Sneak circuit through rectifier and main transformer

Figure 9.25 shows how a rectified supply may act as a short circuit across a brake solenoid and hold the brake off for an appreciable time after the supply contactor has opened. This has an important bearing on the braking of heavy machinery, such as various types of rolls, and has led to accidents in, for example, the rubber industry. This may be prevented by using the connections in Figure 9.26. This is incorporated in Figure 9.37.

Figure 9.27 shows how a fault in a landing push button in conjunction with a full wave rectifier allowed a lift cage to move with the gates open and cause a serious accident.

9.9 Interlocks and guards

H.A. Hepburn[10] defined the various classes of risks. These are listed in Table 9.1.

Table 9.1
PRINCIPLES OF GUARDING (TYPES OF RISKS AND GUARDS)

Type	*Examples*
Revolving shafts, couplings, spindles, mandrels and bars.	Line shafting and countershafts, drill spindles and attachments, boring bars and stock bars.
In-running nips between pairs of revolving parts.	Gear wheels, friction wheels, calendar bowls, mangle rolls, metal-manufacturing rolls and rubber-breaking-and-mixing rolls.
In-running nips of the belt-and-pulley type.	Plain, flanged or grooved belts and pulleys, chain and sprocket gears, conveyor belts and pulleys and metal-coiling apparatus.
Nips between connecting rods or links, and rotating wheels, cranks or discs.	Side motions of certain flat-bed printing machines, jacquard and other automatic looms and various other machines.
Nips between reciprocating and fixed parts, other than tools and dies.	Metal-planer reversing stops, sliding tables and fixtures, cotton-spinning mule carriages and back stops, pillars, etc., and shaping-machine tables and fixtures; tool-steady guides and steady arms on turret lathes.
Nips between revolving control-handles and fixed parts.	Traverse-gear handles of lathes and milling machines.
Nips between revolving wheels or cylinders, and pans or tables.	Sand mixers, edge runners, crushing and incorporating mills, dough brakes, mortar mills and leather-curing machines.

Table 9.1 *(continued)*

Projections on revolving parts	Key-heads, set screws, cotter pins and coupling bolts.
Revolving open-arm pulleys and other discontinuous rotating parts.	Pulleys, fan blades and spur gear-wheels.
Revolving beaters, spiked cylinders and drums.	Scutchers, rag-flock teasers, cotton openers and laundry washers.
Revolving mixer arms in casings.	Dough mixers and rubber-solution mixers.
Revolving worms and spirals in casings.	Meat mincers, rubber extruders and spiral conveyors.
Revolving high-speed cages in casings.	Hydro-extractors and centrifuges.
Revolving cutting tools.	Circular saws, milling cutters, circular shears, wood slicers and chaff cutters.
Reciprocating tools and dies.	Power presses, drop stamps, relief-stamping presses, hydraulic and pneumatic presses, bending brakes and revolution presses.
Reciprocating knives and saws.	Metal-, rubber- and paper-cutting guillotines, trimmers and perforators.
Platen motions.	Letterpress printing machines, paper and cardboard cutters.
Projecting belt-fasteners and fast-running belts.	Bolt-and-nut fasteners, wire-pin fasteners, woodworking machinery belts, centrifuge belts and textile-machinery side belting.
Pawl-and-notched-wheel devices for intermittent-feed motions.	Planer-tool feed motions and power-press dial-feed tables.
Abrasive wheels.	Manufactured wheels and natural sandstones.
Moving balance-weights and deadweights.	Hydraulic accumulators and counter-balance weights on large slotting machines.
Nips between travelling and fixed parts.	Travelling conveyor hoppers and tipping cams, bars or other fixed parts; inclined bucket conveyors and fixed parts.

The following classification of types of guard is also by H.A. Hepburn, but the interpretation is mine and for that I must assume responsibility.

Fixed guards, by virtue of their position, prevent access to the dangerous parts.

Automatic guards move in advance of each operation or stroke of a machine and sweep or push the operator's arm or person out of the way before the stroke is made, or automatically take up a position similar to a fixed guard before the danger can arise. They are sometimes used where material or blanks are fed by hand into the machine. They are best combined with trip guards, but are largely being replaced by interlocked guards.

Interlocked guards must be in correct adjustment before any potentially dangerous operation can commence.

Trip guards stop the machine or automatically cause other appropriate action to be taken immediately danger arises. They can often be combined with advantage with automatic guards or with positional and distance guards.

Positional and distance guards are often a rather elementary form of fixed guard, so placed as to keep an operator or other person at a safe distance from a machine which cannot be enclosed because of the nature of the material being handled or the work being done.

Light beams and electrical induction must now be added to this list, particularly the former, as 'curtains' of interlaced light beams are extensively used on such machines as power presses, bending brakes and guillotines, where the size and shape of the workpieces prevent complete enclosure.

9.9.1 Use of interlocked guards

An interlocked guard must remain locked in position so long as the danger persists, and when it is open or removed it must be impossible for danger to arise; such guards may be divided into two main categories: those in which the dangerous parts of the machine must be at rest before the guard can be opened, and those in which the action of opening the guard stops the machine or the dangerous parts of it.

In both cases it should be impossible to restart the machine until the guard is again in position and locked. In nearly all circumstances, it is not sufficient merely to cut off the power; it is necessary to bring the dangerous moving parts to rest before anyone can get at them, and this generally entails the use of clutches or brakes or both. Even single-stroke machines may need to be braked or scotched to prevent unexpected 'repeat' strokes due to malfunction, and brakes are required wherever a machine can come to rest in an unstable position from which it may move (under gravity or otherwise), particularly if the moving parts are heavy.

Guards, so arranged that the dangerous parts of the machine must be at rest before they can be moved, are always preferable to those where the action of opening the guard stops the machine — unless there is some very good reason to choose the latter — because the electrical and mechanical requirements are simpler and impose less stress on the machine.

9.9.2 Use of trip guards

Trip guards operate only when danger is imminent, and a very quick stop is therefore necessary. Brakes and/or clutches are almost invariably required, but it may not be easy to stop a heavy machine in time to prevent an accident and a trip guard may only reduce its severity. Where interlocked guards can be used these are preferable with trip guards as a last resort.

Light beams associated with photo-electric detector cells are usually of this type (see sections 9.9.15 and 9.9.16).

9.9.3 Operations with guards removed

In many cases, certain work — such as setting-up and adjusting a machine, clearing scrap or waste material or broken and defective products, or replacing damaged parts of the machine itself — can be carried out only when the guards are removed. For these purposes, it may be necessary to allow the machine to perform some part of its cycle of operations whilst unguarded.

Barring gear can sometimes be used, but this often requires very heavy manual labour and it has its own dangers; the only solution may be to run the machine very slowly under careful control. Where electric interlocking is provided, it should be possible to arrange the interlock to prevent normal running but permit the use of a special inching or crawling controller (see section 9.9.10). Such work must always be done by specially qualified persons, and it is easy to arrange that this controller can only be brought into operation by the use of a special key which is in the charge of a responsible official. The location of such a controller must depend on the risk involved, the permitted speed of operation and the rapidity with which the machines can be stopped or the operator got out of the way in the case of a mistake or an error of judgment (see section 9.9.12).

9.9.4 Reliability — 'failure to safety'

It should not be necessary to point out that any safety device should be both as reliable and as nearly immune from breakdown or interference as possible.

In electrical equipment there are special hazards arising from cessation of the supply, from deliberate or accidental open-circuits and from short-circuits. These matters are usually covered under the heading of 'failure to safety', which is generally obtained by ensuring

that, in the case of any foreseeable misadventures, the device shall always do the right thing or nothing at all, and/or if it ceases to function the 'signals shall go to danger' and the whole apparatus immediately assume a 'safe' condition.

9.9.5 Direct mechanical interlocking of guards

For fixed guards, the simplest arrangement is to provide a direct mechanical interlock with a simple switch which controls the motors,

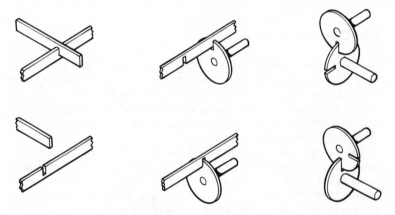

Figure 9.28 Mechanical interlocks between switches and guards

ensuring that power must be cut off before the guard is opened and the guard is closed before power can be restored; i.e. a two-way interlock is required. A number of methods of achieving this are illustrated in Figures 9.28 to 9.33.

9.9.6 Finger nips

Declutching and braking are even more important, and time factors more critical, with trip guards than they are with automatic guards, and there are a number of structural features which merit attention. Such guards should preferably be 'finger-light', and operated by merely a touch. When they have to move links to operate clutches (or even switches), they often tend to be very stiff and capable of inflicting at least a nasty bruise, even though they may prevent worse injury.

A sensitive trip-guard may be useful when complete guarding is very difficult, if it is made impossible to reach the dangerous parts without

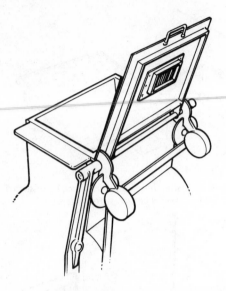

*Figure 9.29 Application of mechanical inter-
locking of switch*

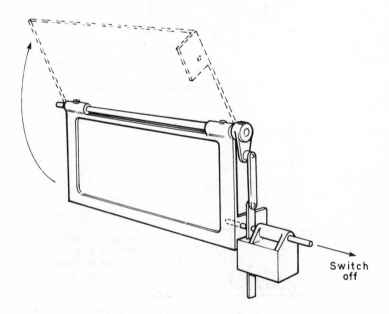

Switch
off

Figure 9.30 Mechanical interlocking of switch

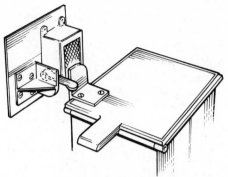

*Figure 9.31 Simple mechanical interlocking of switch.
Guard closed, switch on*

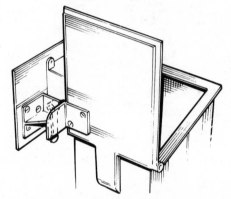

*Figure 9.32 Simple mechanical interlocking of switch.
Guard open, switch off*

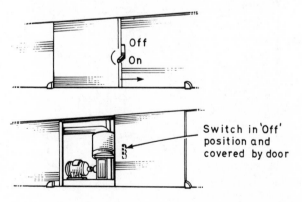

Figure 9.33 Simple interlock between switch and sliding door

touching the trip bar. Operators will soon learn that they must keep clear of the trip bar, and therefore out of harm's way.

A fixed guard may often be arranged so close to an in-running surface that a serious nip is virtually impossible (Figure 9.34), but this is not, as a rule, true. Where trip guards are used in such circumstances care should be taken in their arrangement, bearing in mind that the

Figure 9.34 Arrangement of fixed guard close to an in-running surface

surface or material will move some way at least before it can be stopped and may take the hand or fingers with it. If arranged in a leading position, as shown in Figure 9.35(*a*), it will be dragged down and will grip; if trailing, it will rise a little, see Figure 9.36(*b*).

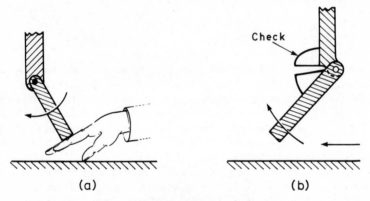

Figure 9.35 Arrangement of trip guards
(a) in a leading position
(b) in a trailing position

In some cases, it seems that instead of a flap, a light roller of tough insulating material might be used (Figure 9.36). The spindle would be a conductor and would rest on two conducting half-bearings at the ends, a *low-voltage* supply passing along it to complete the circuit. If disturbed, however slightly, it would immediately break the circuit at one end of the other; this would cut out all linkages, backlash, etc. It is

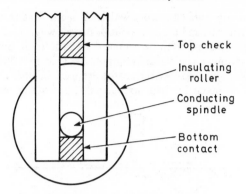

*Figure 9.36 The use of a light insulating roller as a
trip guard*

necessary, of course, to have check stops above each end to prevent it lifting too far and ceasing to be a guard. The voltage may be rather critical, for it must be sufficient to break down any air or dust film but insufficient to cause pitting or electric shock. An alternating supply of not more than 15 V would probably be suitable.

9.9.7 Hinged and rising guards

Some hinged or rising guards must open to insert work, but can be provided with a direct mechanical interlock with the main switch or other source of power arranged so that the moving parts or source of danger are held safe unless and until the guard is closed and locked. This is not always possible, however, because of the type of operation or the shape of the workpiece. In such cases the guard may have to be interlocked with the control circuits.

In other cases where work is fed in manually or must be manipulated while in the machine, a locked guard may be used but there is necessarily a gap and the operator's hand or other parts of his body must be kept out of danger. In such cases trip guards must be used, which may be hinged flaps or aprons, or in some case interlaced light screens, associated with photo-electric cells. This necessarily introduces weak links in the chain of control and should whenever possible be backed up by a second line of defence. All safeguards must always be effective at all times and the machine must not move (or must stop) if any device is out of order or indicates danger (see section 9.9.13).

In some cases automatic or sweep guards are appropriate; these are intended to push the operator, or his hand, clear before the danger can

arise. It is important, however, that they shall be fully effective. In fields surrounded by electric fences the cows quickly learn to feed under the lowest strand of electrified wire without touching it. Similarly, operators soon learn to avoid a sweep or trip guard (the two may be combined) if to touch it would hold up production, or otherwise be inconvenient. Unless a sweep guard is fully effective some nimble operators will learn how to avoid it and adjust work inside the guard at the last minute and may come to grief.

All guards must be mechanically well-designed and robust. Moving or interlocked guards must fail to safety and a very high standard of reliability and immunity to foreseeable failure, interference or damage is essential (see section 9.9.4).

9.9.8 Stopping time

Timing is very important with trip guards and some interlocked guards, on machines which are slow to stop. The following is an example.

In order to prevent accidents with hydro-extractors in laundries and similar places, it is necessary to ensure that the covers cannot be open while the basket is moving. It is important to note that the interlock performs two functions. The cover must be locked in such a manner that the machine cannot be started, whether by electrical or mechanical means such as a clutch, while the lid is open. In addition, there must be means to prevent the cover being opened until the machine is at rest, since it will run for some time under its own momentum.

Considerable care should be taken to see that the mechanical parts are of suitable design and adequate mechanical strength to stand up to the shock of misconceived attempts to open the guard with the machine running. Gate or limit switches may be used to prevent the machine being started with the lid open, and an electrical relay can be devised, depending on the generation of a voltage, to prevent the lid being opened while the cage is rotating.

On one occasion where an electrical device had failed it was found that it depended for its efficiency on very accurate assembly and location. A small error in manufacture or fitting, or alternatively some damage to the structure causing a slight distortion, was sufficient to allow relay contacts to close when they should have been open. A similar point is made elsewhere with regard to certain classes of limit switch. Even if they are manufactured, assembled and installed with a high degree of accuracy, it is likely that they may be disturbed during subsequent repairs by a fitter whose mind is on some other point. Dowel pins may be used to ensure that if a switch or interlock is removed for any reason it can be replaced in exactly the same position.

9.9.9 Trips and stops

It is particularly important with trip guards that once a machine is stopped it should remain stopped until deliberately reset and started. This was illustrated by the following accident. The travelling table of a machine approached so close to a dwarf wall that there was a danger of persons being trapped and squeezed; it was therefore provided with a trip guard in the form of a hinged flap. While a man was standing near the wall, the machine attacked him from the rear and administered a smart slap. This tripped the guard and stopped the machine, but he — naturally — withdrew his undercarriage somewhat; this released the flap, so that the machine inched forward again. This cycle of operations was repeated until he was firmly wedged

Self-resetting safety devices should never be used, whether they are trip-guards, limit switches or emergency-stop buttons. If they operate, there is something wrong which should be investigated.

It is sound policy to arrange matters so that the safety devices cannot be reset by the operator but require the attention of an electrician or fitter.

This feature of ensuring that the machine remains stopped until it is re-set can be obtained by electrical or mechanical means — the latter being preferable; it is illustrated in Figure 9.37. The pushbutton shown is of the latched type and has no return-spring. When the pushbutton is depressed, the contactor circuit is broken and remains broken until the resetting button is pushed, the two buttons being linked mechanically. If one relies on electrical interlocking between the stop and the resetting devices, the machine may suddenly re-start when pressure is removed from the stop button or trip bar, because the resting device has stuck in the 'run' position. The trip bar is also arranged so that the linkage opens the circuit but does not re-close it. Furthermore, when the contactor opens, the low-volt release of the starter opens, and the machine does not restart immediately the trip bar and interlock circuit are reset, the starter needing to be operated in the normal manner. This is desirable, except perhaps when the lifting of a guard is part of the routine operation of a machine, for inserting blanks or similar purposes.

9.9.10 Inching and crawling (see section 9.9.3)

If speed is to be reduced so that certain adjustments can be made with the machine operating in a slow and safe manner, care must be taken to see that it is, in fact, sufficiently slow; it should be a crawling speed and no more. This is fairly easily achieved with d.c. or Ward Leonard units or with brush-shifting a.c motors, but it is more difficult to accomplish

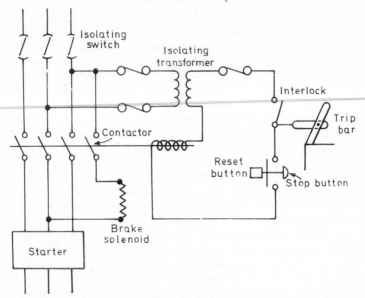

Figure 9.37 The use of an isolating transformer guards against sneak circuits (section 9.8.2). The brake solenoid circuit is operated by contacts on the main contactor so that the brake is not held off temporarily by being short circuited through the motor windings or otherwise (section 9.9.9).

The stop and start buttons are interlocked mechanically so that neither can stick in when the other is operated (section 9.4.2.8). The trip bar opens the interlock but when it is released does not reset it (section 9.9.1). This feature is not easily illustrated. The guard and/or trip bar and/or linkage is robust, rigid and there is no whip or backlash (section 9.9.9), this also cannot be illustrated in a diagram

with induction motors. Where this facility is required, it is best left to the joint efforts of motor and control-gear makers working together. With existing machinery the simplest solution in some cases will be to use a small pony motor for inching and barring.

Far too frequently the speed which can be obtained with the 'crawl' button is much too high, the makers and users relying on the assumption that the device will not be used for a sufficient time for the machine to accelerate far. Buttons may stick, however, and men may fail to leave go; such devices should only be used with a time relay or dashpot which limits the length of time for which the circuit can be held closed.

Preferably, a true inching motion should be used where one pressure of the button moves the machine a definite and small distance forward and then stops it; movement can then be continued only by releasing the button and pressing it again, or by some similar operation.

9.9.11 Methods of braking and stopping

When it is necessary to stop a machine very quickly, some form of braking must be employed, and this will be more effective if the moving parts to be arrested are reduced to a minimum by releasing a clutch. An ordinary friction brake is the simplest type, and it should be held off against the resistance of a spring or counterweight, so that it is applied automatically if the power fails. On the other hand, clutches should be held closed against the spring or counterweight which tends to open them. The reversal of this arrangement caused an accident where a cutter suddenly started to revolve when the power was cut off and continued to run until the motor came to rest.

Electrodynamic braking of various types helps to achieve a quick stop, but is ineffective at slow speeds. It is therefore a useful addition rather than an alternative to a friction brake. Torque reversal (plugging) is very effective, but in many cases a reverse-rotation cut-off will be required, and this is generally an undesirable complication.

Some machines, e.g. hydro-extractors, heavy printing machinery and various forms of rolls, cannot possibly be stopped quickly without wrecking the machine, and trip guards and interlock guards wherein the action of opening the guard stops the machine are unsuitable. If the guard is such that the dangerous parts of the machine must be at rest before the guard is opened, a delay device or rotation detector becomes necessary.

9.9.12 Time factors

Where a very quick stop is required the guard and control gear must be designed as a whole. All backlash and whip must be eliminated from the mechanical features and all unnecessary relays and contacts from the electrical ones. Each pair of contacts which has to open, and every brake or clutch solenoid which has to release, allows the machine to run a little further.

Care must also be taken with the connections. A d.c. motor will generate a back e.m.f. and hold off shunt brakes until the field has partly vanished; there is also a tendency for an induction motor to act as a generator for a very short interval, and if there is a starting or power-factor condenser it may generate for a considerable period while running down. For this reason, it is best to control the brake and/or clutch by independent contacts on the contactor.

If the interlocking switch can be made to intercept the power supply direct without the use of a relay or trip, this is to be preferred, but this

is usually possible only with interlocks where the machine must be at rest before the guards are opened.

9.9.13 Gate interlocks for danger areas (see also sections 9.9.5 and 9.9.6)

It is often necessary to lock the gates of danger areas to prevent access. This applies to some electrical testing such as transformer and switch-gear test enclosures and to places where certain processes are carried out, such as grinding, sieving or otherwise processing some sensitive explosives and to some chemical processes.

Such locks are frequently interlocked with the electricity supply. This may be done electrically, but frequently keys are used which are trapped in the gate lock when the door is open and in the switch when the power is on.

Sometimes master keys are provided, but there is a temptation for skilled persons to use the master key on all occasions, rather than go through the interlocking routine. This has led to a number of serious, mostly fatal, accidents on electrical switchgear, and it is now the practice in the electric supply industry to provide only one master key for use in emergency or under special conditions; this is generally kept in a locked box with a glass front so that, should it be used, the fact is immediately apparent to those in authority. After one fatal accident it was found that every shift engineer in a power station had made or obtained a master key to the main switchboard.

Two methods of locking off danger areas are described below by way of illustration.

(a) Electrical interlocking of gates and screens

Where electrical testing is carried out in a section of a works which cannot easily be structurally separated from the rest of the premises on account of the equipment that is being handled, high portable screens may be used. Attached to each screen is an insulated conductor with a plug at one end of the screen and a socket at the other, so that when all screens are in position they may be linked together and a circuit completed; this will allow a no-volt no-close circuit-breaker to be operated or, alternatively, will energise a contactor in series with the other control gear. The importance of the use of no-volt no-close circuit-breakers is discussed elsewhere. Where such a scheme is adopted, the control should be from a bridge or gallery from which the whole of the danger area may be seen, so that the main circuit-breaker is not closed until a responsible person has checked that no one has been left inside.

Where there is a risk of contact with high-voltage circuits such as those associated with electrical testing, high-frequency furnaces or electrostatic precipitators, it is preferable to arrange for the dangerous circuits to be earthed before doing any work, as well as ensuring that they are dead and isolated. An earth bar is commonly interlinked with the gate or door lock so that the earth cannot be removed until the door is shut and locked. On a number of occasions, however, it has been found that the mechanism of this interlock has been mechanically unsound.

(b) Mechanical interlocking of gates (see section 9.2.3)

Experience shows that electrical interlocking, particularly with fixed structures such as doors to enclosures, is not so satisfactory as mechanical interlocking, since it is easy for testing and operating staff to defeat the interlocks by either disconnecting or short-circuiting the terminals.

Where fixed gates are used a system of key interlocking is often employed. The doors and locks are so constructed that the key is trapped and cannot be removed unless the door is both shut and locked. It is important that it should not be possible to open the door, shoot the bolt, and then remove the key with the door open.

Where there is a number of entrances to an enclosure a simple method is for a different key to be used for each entrance or inspection opening. All these keys are then collected and inserted in appropriate places in a key block; when all keys are present and turned this releases a master key. The individual keys cannot be released unless the master key is in position and trapped, and the master key cannot be released unless all the individual keys are in position and trapped. This master key must be inserted in an appropriate lock on the main circuit-breaker or isolating switch before the latter can be closed, and is held trapped until the circuit breaker is opened. It is quite clear that this system can be employed for interlocking equipment where the risk is other than electrical. It has, for example, been adapted to the locking of chambers in which explosive substances are being ground.

In some circumstances it is necessary to ensure that the machinery has come to rest before a room or enclosure is entered. This may be done by arranging a magnetic brake on the driving motor, but it is sometimes sufficient to arrange for the interlocking switch to be at a fair distance from the door so that everything will have come to rest before anybody can walk, or indeed run, from one to the other with the key. In many situations both electrical and mechanical interlocks are used so as to have a second line of defence.

9.9.14 Permits to work

In the electrical supply industry the use of the permit-to-work system on live gear is fairly widely adopted, and similar arrangements are used in some chemical works. Briefly the arrangement is that before any men are set to work in a situation which is normally dangerous or could become dangerous, a responsible official, who is generally referred to as the 'authorised person', must ascertain that all safety precautions have been taken. Having done this he issues a written permit on a prescribed form to the senior person concerned in the work, generally the foreman or charge-hand. The contents of this form should be brought to the attention of all workmen, and they should be instructed to refuse to work unless they are satisfied that everything is safe.

The permit should to some extent be an instruction detailing exactly what work is to be carried out, the situation in which work can be carried out safely, any special points of danger in the vicinity and what special safety precautions have been taken. If extra work not covered by the permit becomes necessary, this should be amended or preferably a new permit issued. The importance of insisting that no work shall be done outside the permit has been demonstrated by many accidents, including a large proportion of fatalities. That little bit extra frequently involves taking a risk which is not apparent to the people concerned.

It is important that the permit and instruction shall be very clearly worded and thoroughly understood by all concerned. In the electric supply industry it is not always easy to make out a permit since conditions vary considerably, and it has been necessary to insist that the form contains adequate space for description and diagrams. In general industry, however, it seems probable that the main precautions will be much more stereotyped, and could probably be printed on the form leaving a space for any special precautions required for a particular job. If this procedure is adopted each printed instruction should be initialled by the authorised person as an indication that he has checked it. The person receiving the permit should read it through, and sign a declaration at the bottom that he has done so and understands the instructions it contains. When the work is completed he signs a clearance certificate at the bottom stating that the work is completed and all men withdrawn. Until this clearance certificate is received no safety precautions may be interfered with, and, where possible, all isolating switches, control levers, etc., should be padlocked, the keys preferably being retained by the person holding the permit. It hardly seems necessary to mention that there must be no duplicate keys, but, in fact, a number of accidents have resulted from this cause. In particular, individual members of an extensive key series may become interchangeable after a certain amount of wear.

For small enclosures it is usually possible to provide a direct mechanical interlock between the gate and the main switch or isolator controlling the danger area. For extended areas or where the danger area is remote from the controls, as for high breaking capacity circuit breaker tests, either key interlocking or an interlock via pilot cables may be necessary. In the latter case great care is necessary to ensure and maintain the integrity of the interlock circuit and apparatus. There is a case for using both key and electrical interlocks (belt and braces) but there is an inherent danger of staff disregarding the failure of one and not noting that the other is vulnerable.

9.9.15 Interlaced light beams

Over the years there have been many proposals for using beams or pencils of light, often arranged as a 'curtain', in place of trip guards. One example many years ago was designed to protect a power press. It was so arranged that if the operator's hand approached the danger zone it would interrupt a beam of light, de-activate a photo electric cell and stop the machine. There were two fundamental defects. In the first place it could be defeated by applying an electric handlamp or torch to the photo-electric cell. One method of overcoming this trouble is to use a pulsed or chopped light signal and a tuned relay in the detector circuit which will only accept pulses at one frequency. In the second place, if the press was committed to making a stroke it was easy to push a piece of wood under the descending tool (it is not advisable to risk one's hand).

This latter defect was again a matter of cumulative time delays. The current through the cell was insufficient to operate a contactor, therefore a small telecommunication relay was used to break a contact or coil circuit; this in turn operated a solenoid-controlled valve which released the air pressure in a compressed air system (which takes a measureable time) and this withdrew the clutch and applied the brakes; also as the moving parts had considerable inertia the machine could not be halted instantaneously without damaging it.

There has been considerable pressure, however, to allow the use of photo-electric guards, partly because mechanical guards are heavy and slow down production, but perhaps more important, some machines such as bending brakes are very difficult to guard effectively because of the size and shape of the workpieces and the need to guide and manipulate them manually. This is perhaps more important in jobbing engineering and on small production runs than in mass-production.

This matter has been reviewed by the Joint Standing Committee on 'Safety in the use of Power Presses' who over a number of years have issued a series of reports, published by HMSO[11].

9.9.16 Electronic and solid state controls

When electronic control equipment was first introduced into industry on a substantial scale, between 1945 and 1950, its designers adopted research laboratory and telecommunication practice with the result that some of it was inadequately protected and was not sufficiently robust to withstand industrial conditions.

A control unit in a steelworks the size of a cathedral need not fit into a biscuit tin when greater robustness and better clearances and creepage distances could be achieved if it was as large as a filing cabinet; it may be bumped by a crane hook or suffer from a deposit of semi-conducting mill scale, or from sulphur-dioxide in the air. Not all items can be put in air conditioned enclosures. Further, where plant may cost several thousand pounds a minute to operate, reliability is the first requirement and components should be under run and over insulated if that will prolong their lives. There should be no attempt to get a quart into a pint pot to save a few pounds.

Thermionic valve circuits are primarily high impedance systems and therefore it is virtually impossible to detect earth leakage or indeed leakage between items as can be done with conventional apparatus, leakage on, say, a grid bias can silence an alarm or cause loss of control and/or immediate danger. This can be overcome to some extent by the use of cathode follower devices, but not entirely.

Solid state devices such as transistors are primarily low voltage low impedance systems and are easier to protect. There is little tendency to tracking and sparking or arcing are not serious hazards. With reasonable clearances and creepage distances, leakage can be eliminated but, in industrial situations the possibility of surface contamination must always be considered.

On the other hand, such devices are very sensitive to stray inductance and can be irretrievably damaged by moderate over-voltage and over-current. Attention must therefore be paid to the voltage and current withstand values discussed in Chapters 6 and 7. In the absence of special protection stray voltage spikes and potential gradients, spilling over from power circuits can cause havoc. On one occasion a transformer failure in a 'grid' substation caused an earth potential difference of several thousand volts between the two ends of a control cable and current flowing along the lead sheath led to arcing. The cable end was destroyed and control lost. This would have had disastrous effects on semi-conductor units if they had been in use. This sensitivity also introduces difficulties in testing.

Protection against over-current can be provided by taking advantage of the low let-through values of $[I^2 t]$ of specially designed fuses (see Chapter 7). Protection against voltage spikes is commonly given by

shunt connected Zener diodes, as for intrinsically safe equipment. The maximum electrical and spacial segregation of circuits should be provided and as an alternative to isolating transformers and other coupling units, increasing use will be made of optical coupling in which there is no electrical connection between one part of a device and another, the necessary signal being communicated by a light pulse. This can be achieved over considerable distances and round corners by fibre optics and relays operated by light emission diodes are becoming available, but it will be necessary to ensure that the physical separation of the two circuits is sufficient and of adequate breakdown strength to stand up to service hazards in industry.

9.10 Control logic

Since 1945 various systems of switching have been developed based, rather indirectly, on Bool's Mathematical Analysis of Logic of 1847 and later writings. If, for example, one wished a controller to carry out the programme 'If lift gates A, B and C are closed, the cage D is not overloaded and control button E is depressed then the lift will move to floor E', or symbolically:

If (A *and* B *and* C *but not* D) *and* E then . . . .

this could be realised by the circuit in Figure 9.38 where the switches are shown in the unoperated state. (*Note.* A useful convention in drawing control diagrams is to place normally open contacts with the bar above or to the right of the line and normally closed contacts with the bar below or to the left, see Figure 9.38. A mistake or drawn placed bar is not then misinterpreted.)

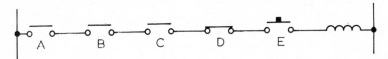

Figure 9.38

Another example could be:

If the safety cage is closed *and* the temperature is correct *and* the *start* button is pressed *then* . . . .

or

If the stop button is pressed *or* the safety cage is open *or* the temperature has fallen *then* . . . .

The terms AND, OR, NOT, are called logic units. Others used are OR ELSE, NAND (not and) and so on; different manufacturers build up their systems from slightly different units, but today relays are being rapidly replaced by assemblies of solid state units such as transducers, diodes and transductors. Quite complicated logical 'bricks' may be built up on small unit 'chips' by specialist sub-contractors and are obtainable 'off the shelf'.

The study of the safety of such control systems must follow different lines to that for older equipment and practice. The development of new devices has recently been so rapid that no general recommendations can yet be made but these devices are extremely reliable.

Individual control manufacturers have developed their own systems for checking new equipment and locating faults in operation, which is facilitated by the fact that the output at any point is usually two valued (0 or 1, + or −).

The upper limit of development is probably the use of 'on line' minicomputers, which may be duplicated as a check on each other. Built-in self-monitoring and checking procedures have also been developed which regularly check the overall response of functional units, but not individual components. All this, however, entails the employment of expert technical staff to service such apparatus.

9.11 Assessing control systems

It is of the utmost importance that designers of interlocks and control systems should study how they may be defeated by the development of sneak circuits (as described in section 9.8.1) and also as a result of accidental or careless operation. On one occasion a girl lost two fingers on a guillotine because she unconsciously pressed two buttons in the wrong order, and it was subsequently found that if a particular split pin failed the machine could only be stopped by pressing the start button. This was the result of an obscure feature of an otherwise satisfactory design.

Erectors and maintenance men almost invariably know how to defeat any system because they cannot do their job without doing so and it is always worth having a friendly chat with them. But because they find it necessary, provision should be made in the design for this to be done in an authorised and controlled manner and for ensuring that the safeguards must be restored before equipment can be put back into service. This has been covered by some British Standards on switches and controllers. Maintenance fitters often need to watch

contactors etc. in operation, and windows should be provided for this purpose.

With the increasing sophistication and the introduction of logic units it has become difficult to foresee all the ways in which danger may arise. At the same time HM Electrical Inspectors have been asked to vet larger numbers of proposed designs, particularly for photo-electric guards and controls on power presses and bending brakes. In the last few years my former colleague Mr. Goodman developed a method of dealing with these matters, which is given in outline below.

9.11.1 Procedure

1. Redraw maker's diagram using a familiar convention and correct obvious mistakes. In the process one learns how the system works and some potential weaknesses become apparent.
2. Reduce diagram to a standard form suitable for simulating on a 'plug board' with standardised components. It may be necessary to substitute equivalent components or sub-circuits for some functions, particularly with solid state logic controls. The results of testing will require re-interpretation in terms of the actual scheme.
3. Run off a number of copies (see (b) below).
4. From this the normal operation can be checked.
5. Follow by checking the result of foreseeable abuse, e.g. push-buttons operated in the wrong order or at the wrong time.
6. Check the effect of single component faults such as relays stuck open or welded closed; cams operating switches worn or incorrectly timed; capacitor, diode and transistor faults, e.g. failing to the open or short circuit condition.
7. Examine the circuit to check where single point faults may result in simultaneous mal-operation of more than one component.
8. Check the effect of foreseeable coincident faults (see Note (c) below).

Notes: (a) Once set up the analysis can be carried out on a simulated system very much more quickly and with greater certainty than by tracing faults on a complex diagram.

(b) Each step and its result should be recorded on a separate photocopy of the diagram and checked to see that no error has been introduced by the simulation technique. The state of each relay or logic function under fault conditions can be checked against the normal state at every stage of the operating cycle.

(c) With a complicated control system, particularly if (8) is to be attempted, the procedure may still be exacting and time-consuming

and the simulation can become very complicated and confusing. Consideration must then be given to computer-aided methods of analysis. Once accomplished these can tabulate in a few seconds all possible permutations and combinations of errors, which would take many hours by other methods. But the reliability of a computer cannot exceed that of the programme and important conclusions should be rechecked on the diagram or simulator.

This is, however, only half the problem; it is necessary to assess the electrical and mechanical reliability of the components which are to be used, and finally the apparatus itself when it is made. Defects are likely to be found which were unlikely to be appreciated on drawings and diagrams.

9.12 Control centres — layout and instruments

Centralised control has always had a place in the electrical supply industry and has been used in the gas industry, for fuel pumping and water supply and for large process plants whose operations had necessarily to be co-ordinated. Where the distances are considerable control must be by electrical means, which could be by pilot cables, or telephone or radio messages. In due course laser beams and fibre optics may be used for this purpose.

Three fundamental requirements apply in all cases. 'Signals' must be transmitted, received and interpreted correctly. Relevant information must be displayed unambiguously giving the present state of the system and calling attention to important changes. The operator's work load and reasonable comfort must be studied.

Leaving aside the transmission of information, or signals for later consideration, first priority should be given to the display of information. Anyone who has seen a power station control room or the cockpit of a modern jet aeroplane must wonder, 'are all your instruments really necessary?' Assuming that they are, are they equally important? Could some of the information be recorded automatically, with, if necessary, arrangements for instant recall and display when required?

Assuming this has been settled, is the information displayed in the best possible way? A digital clock may be the best for reading the time *accurately* but a conventional clock could be better for a hurried glance, particularly from a distance; it is not necessary for there to be any figures on the face, in fact the position of the hands alone may be quite enough. There are occasions when accuracy is important and others when quick response is essential; this also applies to all gauges and instruments.

Another matter of importance is the lettering, including figures. This was discussed by Golds[12] in the Conference on Electrical Instrument Design in 1951 and Shackel[13] in 1961, together with the best form of scales for different purposes[14]. Perhaps the most important general conclusion that can be reached is that instrument faces should be clear, uncluttered and not ambiguous (the need for clarity is illustrated by air disasters caused by mistaken reading of altimeters). Tabulating accurate data is best done automatically, while displays should concentrate first on essential information for dealing promptly with an emergency (this *may* not entail high accuracy) and only secondly on information required for routine operation, e.g. for trimming controls for optimum conditions.

There are great advantages in making displays pictorially suggestive, thus dials may be arranged so that the pointers are parallel when everything is normal, so that any one out of line is immediately obvious. Wattmeters or ammeters may be arranged with centre zero to indicate the direction of flow at a glance. On an industrial control board the operation of the plant can be indicated by formalised pictures of the equipment. This is important for semi-skilled operators or situations where displays are not under continuous observation and quick decision may be necessary by staff who may not be very familiar with a particular display.

Other matters of importance are illumination levels and the elimination of confusing reflections, e.g. of lights or windows, and the distance and angle from which indicators must be viewed. An interesting historical study of such problems is *Planned Seeing* by Bartlett and Mackworth of the Medical Research Council's Applied Psychology Unit at Cambridge[15] which deals with visibility in Fighter Command control rooms during the 1939–45 War (including among other matters the legibility of letters and numbers).

The temperature, illumination and humidity will also affect the performance of control staff and such matters as toilet facilities and the ability to make a cup of tea in quiet moments are important. Good control engineers need a rather special temperament: long hours of quiet routine must not lead to boredom but in an emergency they must be quick, decisive and unflappable.

The Applied Psychology Unit of the MRC have carried out valuable studies of the work load of persons who need to be continually alert and vigilant over extended periods, which includes car drivers and airport control operators. There is evidence that men with experience can carry on for long periods under normal conditions without their skill being notably reduced by fatigue, but they lose the 'spare capacity' to deal with a flood of extra 'information' in an emergency. The effects of moderate amounts of alcohol are similar, and no one can continue to

give very close attention to a changing display for more than a short time, without missing some changes. That is the reason why alarm signals are needed (and for the cup of tea).

In a control room there may be many telephones as well as alarm bells. Those providing vital information in an emergency should have distinctive notes and it may be advisable to have some means of ensuring precedence for certain information and perhaps suppressing some incoming calls at busy times.

9.13 Remote and supervisory instrumentation and control [16]

Supervisory control and modern process control is usually carried out at low voltages by means of standard telecommunication equipment or logic elements, and may include online computers.

Primarily, it is necessary to transmit accurate information to a central point without delay and without error, and, on the basis of the information received, to send out automatic operating instructions. In the following general recommendations central control is assumed to be personal, but, if an online computer is in charge, the necessary alteration to the statements can easily be made:

(a) Equipment must fail to safety; i.e.

(i) Every remote indicator must give the correct indication, or no indication.

(ii) Every operating handle (device) must cause the correct operation, or no operation.

(iii) Every interlock must allow only safe operation, or prevent all operation.

(iv) Any failure to function correctly must be made immediately apparent.

(b) (i) Every indicator or interlock should span the complete equipment and operation in time and space.

(ii) Every important indication should be self-checking, and it is an advantage to include the operator in the check.

(iii) Important incoming signals should normally take precedence outgoing instructions, where the same channel or terminal equipment is used, so that action may be based on the latest information.

The meaning of set (a) is self-evident, but (b) (i) requires some further explanation. The intention is that any signal giving, say, the position of a switch or a valve should be actuated from the link nearest to the point of action. Thus, when the semaphore signal on a power-station control board was actuated by a mechanical link several steps back from the

contacts, it failed to show that an intermediate link had fractured, and an isolator shown as open was, in fact, closed. It is also necessary for the latest possible information to be available and displayed.

Point (*b*) (ii) refers primarily to the transmission of information. When the operation and the control centre are remote from one another, a signal may be missed, wrongly received, or misinterpreted. It should therefore be checked, automatically if possible. There are, in principle, two ways of doing this; the methods of encoding the signal can, within limits, be made self-checking, or the signal can be registered, transmitted back, and compared with what was sent out (just as a telephoned message can be read back). The former is the quicker, but the latter is more certain.

Spurious signals over substantial distances may be caused by electromagnetic or electrostatic induction, particularly where pilot conductors follow the same route as power conductors. The importance of the very large differences in earth potential which may occur during major power faults is sometimes overlooked, but they can put a communication channel permanently out of action. Failure of a communication channel should raise the alarm and often initiate safety action; e.g. 'put all signals to danger'.

9.14 Circuit-breaker selection and operation

When one wishes to operate a circuit-breaker remotely it is important to ensure that the correct switch has been selected and that the operation has actually been carried out. The following description incorporates details from existing equipment, but all the features listed may not occur in any one installation.

The action of turning any key on the switch panel lights a corresponding lamp supplied from the switch-operating circuit. If this is placed alongside the switch symbol on the mimic diagram, when this is separated from the control desk, it will perform the joint function of proving the circuit and ensuring that the operator is at the correct panel, unless he is very unobservant. If he then depresses the 'operate' button, the fact that the correct operation has been successfully carried out will be shown by a change from red to green (or vice versa) and the illumination of a yellow light will indicate the change of conditions.

9.14.1 Circuit proving

This example illustrates in an elementary way two principles which become of major importance when supervision is exercised over considerable distances, i.e.

1. Proving the selection before carrying out any operation.
2. Closed-circuit checking, via operator–control panel–operating circuits–switch–indicator pilots–mimic diagram–operator.

9.14.2 Position indicators

Remote automatic indication of the position of isolators has not been so common, but it is becoming important with the introduction of motor-operated isolators in large switchhouses. Figure 9.39 illustrates a

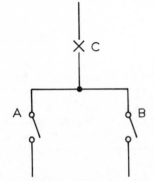

Figure 9.39 Switch and isolators for busbar equipment

simple case of the isolators for a duplicate busbar equipment. It was important that isolators A and B should not both be opened if C were closed, since this might result in operating an isolator on load. This was arranged by providing remote indication of the isolator positions on the mimic diagram and also by providing an interlock to prevent A (or B) opening if C were closed and B (or A) were already open, which was effected by auxiliary contacts operated from the isolator operating rods.

As initially arranged, the indicator moved from red to green whilst the isolator was partly open and the contacts only slightly separated, which would be dangerous in the event of the operating supply failing or the gear jamming, for it would indicate a switch as open when in fact it was nearly closed. It should have been so arranged that in the half-open position the indicator showed neither red nor green. Secondly, the interlock became effective only when both A and B were fully open, so that it was quite possible to open both isolators at once (but not consecutively). This was an instance of failure due to the indicators and interlock not spanning completely both the equipment and the operating sequence.

Two further points should be mentioned here. It is highly desirable that the relay circuits should be so arranged that indication of circuit-breaker operation automatically takes precedence over all other features. This is particularly important during emergencies. It is further important that, where indications of changes are automatically initiated from the substation, the control engineer should have a check key. The board will always show the conditions at the time the last message was received, but communication with the substation may have been subsequently lost. Depression of the check key should extinguish all lamps representing the selected substation and should no reply be received they will remain extinguished. The board will, in effect, reply 'I don't know'.

9.15 Railway signalling

Few pieces of machinery are more dangerous if they get out of control than railway trains or locomotives. From the earliest days railway-signalling systems have been developed to prevent accidents, so far as seemed humanly practicable, and although these systems fall outside the scope of this chapter a brief outline of one of the early schemes of automatic operation is included, since it illustrates a number of the principles described in the preceding sections.

An early installation at Doncaster was typical of many important junctions, it was necessary to route traffic through a maze of 'roads' and branches and to ensure that any route selected was clear of traffic, that the points were correctly set and that the route selected did not conflict with any other route in use or about to be used.

Predetermined routes were selected by a number of miniature motor-driven drum controllers. Each position of a drum corresponding to one route, the cams and fingers making all (or nearly all) the necessary connections for carrying out the operation. It is clear that, if the controller came to rest at any one route position, no conflicting routes on the same drum could be set up at the same time. Routes which might be needed at the same time were allocated to different drums, but it might happen that, say, routes A and B on different drums did not conflict with each other but they both conflicted with a third route C, and resort must then be made to electrical interlocking, special fingers on the drums being provided for this purpose. All interlocking of this type is on the break-circuit principle, to ensure failure to safety.

The interlocking arrangements may be approximately summarised as follows (the actual details are more complex):

The signalman selects a route by turning the appropriate knob (route-selection switch) on a mimic diagram which is very similar in

layout and construction to a power-supply mimic diagram. Then, provided that

(*a*) no conflicting route has been set up or is in the process of being set up,
(*b*) all conflicting signals are at 'danger', and
(*c*) no vehicle is already on any part of the proposed route or
(*d*) is approaching on a conflicting route and is within the maximum braking distance,

a clutch engages and the appropriate drum starts to revolve. Such is the method of preventing conflicting routes being set up. The detection of standing or approaching trains and wagons is achieved by track-circuiting, which is standard railway-signalling practice.

When the drum reaches the desired position, it is mechanically locked by a relay and clutch. A proving circuit is then automatically established only if the drum

(*e*) is locked,
(*f*) does not conflict with the setting of any other switch, and
(*g*) corresponds with the position of the selecting switch on the control panel.

This allows the necessary power relays to set the point-operating machines in motion.

When the points have moved, further proving circuits check that

(*h*) all points are in positions corresponding to the route selector switch and are mechanically locked,
(*i*) special measuring rods show facing points to be within approximately one-sixteenth inch of the correct position, and
(*j*) the signals on all conflicting routes are at 'danger'.

The signals could now be set and the route made available for traffic. A similar careful series of checks applies in 'breaking down' any selected route to leave the controls ready for the next operation.

It will be seen that, in a very complicated problem, closed-circuit checking of response and instruction is carried out at every stage of the operation and the check very carefully spans the complete operation, since a clear road is indicated only when an actual measurement shows that every important point is accurately closed in its correct position and mechanically locked. This was an improvement on much electrical switchgear practice.

9.16 Reliable transmission of information

It is essential that any remote supervisory system shall be reliable and, in particular, that danger shall not arise from an incorrect indication,

such as a switch being indicated as open when it is, in fact, closed, or from an incorrect operation, such as the closing of the wrong switch. To ensure that this shall not occur various self-checking devices have been adopted.

Remotely controlled and supervised, unattended substations have existed for at least 50 years; for example in one large city all the rotary converter substations for the tramways were of this type, and during the 1939/45 war the power station turbine and boilerhouse in a frequently bombed town could be left unattended during severe raids and controlled automatically, with remote supervision from a relatively safe place.

Effective 'Remote Supervisory Control' of a large electrical supply network or process plant requires the accurate transmission of data and instructions over considerable distances, which on a supply network could be 50 or 100 miles. To lay special pilot cables for each function becomes excessively expensive and originally automatic telephone equipment was adapted for this purpose.

The first step was to use a telephone dial to transmit signals which selected instrument circuits at the far end, which would transmit data to the control centre relaying voltages, load, circuit breaker position (open or closed) etc, and would transmit 'orders' to open or close switches, read instruments, change transformer taps etc.

Three requirements were clearly necessary (a) the information received must be accurate, (b) the orders must be correctly carried out and (c) there must be certain 'priorities' in the use of the line. This was solved by arranging that every instruction must be *automatically* repeated back and confirmed before any action could be taken. This stage of development was successful but too slow for large systems.

The next stage was for the message to be stepped out on a rotary switch (uniselector), impulses from which stepped round a corresponding selector at the far end and they automatically sent impulses on a separate channel to step round another selector at the first transmitter, and the system locked out if its position did not correspond to that of the selector which initiated the action; thus establishing a closed loop check. This proved reliable but was still slow, and required the duplication of channels.

A later development was not to repeat back the impulses but to count them. This requires impulses of two types (+) or (−), long or short, arranged so that the total number remains constant. For example for a total of 25 one could have say 13 positive followed by 7 negative and a lost or spurious pulse would cause the apparatus to lock out.

This, however, does not make the best use of the facilities and a mixed signal of alternative pulses, similar to morse code may be used; this leads to binary digital coding and the use of self checking codes.

The whole safety of such a system depends on the method of checking coded messages for errors.

9.17 Other developments. Remote control

At the time my report was prepared on which the above account is based, Shannon and Weaver's book[17] on the 'Mathematical Theory of Communications' had just been published (1949), and the opportunity was taken to show the relation between his work and possible developments in supervisory control. In fact my paper of 1952 was the first to do so and to indicate how efficiency and reliability of supervision could be improved by coding methods[16].

All that need be said here is that signalling on a binary code may be compared with a game of 'twenty questions' to which answers may only be yes or no. The possible number of 'objects' which can be distinguished in this way would be 2^{20} or about one million if played systematically. A five (binary) digit code has 32 values (sufficient to identify any letter of the alphabet) and extra digits (redundancies) can be used to identify errors in transmission. For example a count will establish if a pulse has been missed or a spurious one picked up, and if the binary values are interpreted as plus or minus one extra pulse will ensure that their product is always positive and thus detect any *one* wrong digit. Since that time many more sophisticated methods of coding have been devised, and with modern high-speed computer units there is practically no limit to what can be done — if it is worthwhile.

I feel, however, that many engineers may prefer closed loop systems equivalent to those discussed above since the whole process can now be carried out in micro seconds and the loop test is easily understood, which some codes are not, and can be made very searching (e.g. by including the action of the operator (or computer) in the loop).

9.18 Closed loop feed-back control of machinery or processes

I have not dealt with the stability of conventional closed loop controls because instability has not, so far as I am aware, led to serious accidents. There is always a possibility for example that an open-circuited loop in an error actuated speed control could cause a machine to accelerate indefinitely and burst, as a shunt d.c. motor will do with an open circuited field coil. On many systems this is prevented by the non linear characteristics of the control, but it is best to use overall monitoring and protection rather than adapt the basic control system.

Designers should however be well aware of the work of Nyquist and Bode and there are many standard works on this aspect of control.

Acknowledgement

This chapter has been based partly on a series of articles on the *Electrical Times* and papers published in the *Proceedings of the Institution of Electrical Engineers*. Parts of the Safety Health & Welfare Pamphlet *Electric limit switches and their applications* have also been used as a basis for this chapter with the permission of the Controller of Her Majesty's Stationery Office.

References

1. BS 2631: 1955: *Oil switches for alternating current systems*, British Standards Institution
2. For further information on 'over-current protection', see Chapter 7
3. BS 816: 1951, Pt 1: *Requirements of electrical accessories*, British Standards Institution
4. BS 3185: 1972: *Heavy-duty composite units of air-break switches and fuses for voltages not exceeding 660 V*, British Standards Institution
5. Electricity (Factories Act) Special Regs. Nos 7, 11 and 22, HMSO
6. Health & Safety at Work, Booklet 24, *Limit switches and their applications*, HMSO
7. W. Fordham Cooper, 'Electrical control of dangerous machinery and processes', Parts 1 & 2, *Proc IEE* (1947, 1951/2)
8. Thornton, W.M., 'The microgap switch', *Jour IEE* 80 (1937)
9. Taylor, J. Russell and Randall, C.E., 'Voltage surges caused by contactor coils', *Jour IEE* 70 (1943)
10. Hepburn, H.A., 'The fencing of dangerous parts of machinery'. *Paper read to Inst of Mech E* (1948/9)
11. Reports of the Joint Standing Committee on 'The safety of power presses', HM Inspector of Factories, HMSO
12. Golds, H., 'Figures on gauges and instruments', *Discussion paper at the Conference on Electrical Instrument Design* (1961)
13. Shackel, B., 'Ergonomics in instrument design', *Instrument Practice* (June 1961)
14. 'Design of scales for industrial instruments', *British Jour of Applied Physics* *12* (1961)
15. Bartlett, Sir Frederick and Macworth, Dr. N.H., *Planned Seeing*, Air Ministry Publication 3139B, HMSO (1950)
16. Fordham Cooper, W., 'Remote and supervisory control', *Proc IEE* 99 Pt II (1952)
17. Shannon, C.E. and Weaver, W., *The Mathematical Theory of Communication* (1949). This reprint's Shannon's original paper with a reprint of an appreciation or assessment by Warren Weaver.

Chapter 10

Fire and Explosion Hazards – General Principles

10.1 Introduction

Electricity is not alarming and, with care, its use can lead to greatly increased safety. It has far too often been made a scapegoat when no other cause for a fire could be thought of. Fire reports are sometimes prepared by persons with a great knowledge of fires but a limited knowledge of electricity, and, on occasion, it has been possible to prove conclusively that an assumed electrical cause was invalid. A great deal is known, however, about electrical fire hazards, which are very diverse, and all possible use should be made of that knowledge.

There have been several statistical studies of fire hazards, among which those by Lawson and Fry[1], and by Gosland[2] may be mentioned. It will be noted that the great majority of fires and fatalities occur in domestic and similar premises, although the most serious losses of life have occurred in premises such as stores, hotels, and places of entertainment, where many people may be at risk. Some major fires have, however, occurred in manufacture and caused loss of life because of particular hazards such as large amounts of inflammable liquids.

In this chapter, there is a general review of the scientific bases of fire hazards and their prevention, followed by an account of the particular problems of danger areas and some notes on miscellaneous special hazards and premises. Some possible fire hazards associated with transformers, switchgear etc, are dealt with in Chapters 9 and 11.

A works electrical engineer or an engineering consultant will usually be held responsible for the safety of his equipment in the conditions which can arise on site. It is important, therefore, that he should be able to recognise potential fire hazards. Nothing will, however, be

229

gained by installing a great deal of expensive flameproof or other special equipment which could have been avoided by relatively simple precautions in planning the layout of other plant and buildings. It is probable that the engineer will be blamed for not suggesting these changes, even though he is not an expert and their implementation is the duty of other persons. But he must be careful not to accept responsibility for work on which he is not an expert.

For this reason the following discussion commences with a moderately detailed account of fire hazards in general, which is followed by electrical precautions in particular. It is not to be expected that easy cut-and-dried solutions are available, except in simple cases. Hazards are constantly increasing and changing on the one hand while, knowledge increases and practice improves on the other. In this chapter I have tried to provide the background information, but the actual choice of precautions for specific purposes must be made by the expert on the spot. A number of aspects of these matters are subjects of acute controversy at the present time and I do not claim to be equally expert on all of them, but I have endeavoured to suggest promising paths through the jungle.

10.1.1 The central issue

Electrical installations may cause fires in three principle ways. Most insulating materials are in some degree flammable, ceramics, minerals such as asbestos or alumina, and mica being the chief exceptions, therefore serious overloading is necessarily a fire risk. Overloading may also set fire to flammable material in the immediate neighbourhood, for example wooden and similar panelling over cable runs has caused disastrous fires in shops, theatres and other entertainment centres.

Sparks and arcing caused by breaks in conductors or short circuits can have similar results. They may also ignite flammable gases, vapours and dusts or powders. Badly designed or inappropriately used electrical heating appliances are also a common source of fires, e.g. radiators, convectors and electric irons.

Protection against these risks is usually simple, except in the case of electrical apparatus for use in danger areas where significant amounts of flammable gases, vapours, dusts or explosives may be present, this falls within the scope of Chapters 6–8. For this reason this chapter deals primarily with the recognition of the hazards, i.e. with the characteristics of flammable material. Chapter 11 discusses the difficult problem of the provision of safe equipment for danger areas and Chapter 12 deals with some specific hazards.

10.2 The ignition of gases and vapours

In a gas mixture only the activated particles will react together. These have received sufficient excess energy by impact to disturb the inner equilibrium of the atoms and valency electrons to make the change possible. Activated 'particles' so formed may be atoms, ions, radicals with unsatisfied valency bonds, or molecules in an unstable condition. This energy comes from the kinetic energy (or heat) of the gas, to which an electric field may contribute, and the proportion of activated particles increases exponentially with temperature, so that a rise of approximately 10°C doubles the speed of many simple reactions. (This is a very brief summary of a complicated process and cannot therefore be completely correct. It is intended only as an introduction to more practical considerations.)

If, in a very small volume of gas mixture, the energy released by exothermal reactions exceeds the energy lost by radiation, conduction or diffusion (and endothermal reactions if there are any), the temperature will rise until a balance is reached, and the reaction will spread, i.e. there will be a small self-propagating and expanding flame. The temperature at which this commences may be called the ignition temperature, although it has no absolute value but depends to some extent on the experimental conditions; the minimum volume which must be ignited for the flame to spread is usually of the order of one cubic millimetre. If the flame is smaller it will use up its fuel and go out.

10.3 Spark ignition

The more quickly energy is supplied and the more it is concentrated in a small volume, less gas will be lost by radiation, etc and it will, therefore, ignite more readily. An electric spark, particularly one caused by discharging a condenser is ideally suited to do this and it is a very efficient means of ignition. This is in fact what happens with a car ignition circuit; when the contact breaker opens, electro-magnetic energy from the coil is transferred as electro-static energy to the condenser, and as the voltage builds up across the gap in the sparking plug, the insulation of the intervening gas breaks down and a very 'hot' spark occurs. (To be more precise there is normally an oscillating relaxation transient, charge being rapidly transferred backwards and forwards across the gap until the energy is expended in the spark or as I^2R loss.) It is a moot point whether the activation energy is supplied as heat alone or whether a substantial part is in the form of ionisation.

The design of intrinsically safe electrical circuits and apparatus is based on the limitation of spark energy when a circuit is ruptured. This is discussed below.

Table 10.1
PROPERTIES OF SOME FLAMMABLE GASES, VAPOURS AND LIQUIDS AND RELATED T CLASS AND APPARATUS SUBGROUPS

Compound	Melting point °C	Boiling point °C	Vapour density compared with air (air = 1)	Flash-point °C	Flammability limits in air				Ignition temp. °C	T class	Apparatus group
					Lower Vol%	Upper Vol%	Lower g/m³	Upper g/m³			
acetaldehyde	−123	20	1.52	−38	4	57	73	1040	140	T4	IIA
acetic acid	17	118	2.07	40	5.4	16	100	430	485	T1	IIA
acetone	−95	56	2.0	−19	2.15	13	60	310	535	T1	IIA
acetylacetone	−23	≈140	3.5	34	1.7	—	—	—	340	T2	IIA
acetyl chloride	−112	51	2.7	4	5.0	—	—	—	390	T2	IIA
acetylene	−81	−84	0.9	—	1.5	100	65	380	305	T2	*†
acrylonitrile	−82	77	1.83	−5	3	17	105	360	480	T1	IIB
allyl chloride	−136	45	2.64	<−20	3.2	11.2	105	—	485	T1	IIA
allylene	−103	−23	1.38	—	1.7	—	28	200	—	—	IIB
ammonia	−78	−33	0.59	—	15	28	105	—	630	T1	IIA
amphetamine	—	200	4.67	<100	—	—	—	—	—	—	IIA
amyl acetate	−78	147	4.48	25	1.0	7.1	60	550	375	T2	IIA
amyl methyl ketone	−35	151	3.94	(49)	—	—	—	—	—	—	IIA
aniline	−6	184	3.22	75	1.2	8.3	—	—	617	T1	IIA
benzene	−6	80	2.7	−11	1.2	8	39	270	560	T1	IIA
Benzaldehyde	−26	179	3.66	65	1.4	—	60	—	190	T4	IIA
benzyl chloride	−39	179	4.36	60	1.2	—	55	—	585	T1	IIA

*Not yet allocated to a subgroup.

†For C_2H_2 and CS_2 flameproof equipment is not specified, but intrinsically safe equipment of group IIC may be used.

This table is reproduced from BS 5345 Pt 1 1976 (Table 5) by permission of The British Standards Institution, 2 Park Street, London W1A 2BS, from whom complete copies can be obtained.

Table 10.1 *(continued)*

Compound	Melting point	Boiling point	Vapour density compared with air (air = 1)	Flash-point	Flammability limits in air				Ignition temp.	T class	Apparatus group
					Lower	Upper	Lower	Upper			
blue water gas	—	—	—	—	—	—	—	—	—	T1	IIC
bromobutane	−112	102	4.72	< 21	2.5	—	230	—	265	T3	IIA
bromoethane	−119	38	3.76	<−20	6.7	11.3	300	510	510	T1	IIA
butadiene	−109	4	1.87	—	2.1	12.5	25	290	430	T2	IIB
butane	−138	−1	2.05	−60	1.5	8.5	37	210	365	T2	IIA
butanol	−89	118	2.55	29	1.7	9.0	43	350	340	T2	IIA
butene	−185	−6	1.94	—	1.6	10	35	235	440	T2	IIB
butyl acetate	−77	127	4.01	22	1.4	8.0	58	360	370	T2	IIA
butylamine	−104	63	2.52	− 9	—	—	—	—	(312)	T2	IIA
butyldigol	−68	238	5.59	78	—	—	—	—	225	T3	IIA
butyl methyl ketone	−56	128	3.46	23	1.2	8	50	330	(530)	(T1)	IIA
butyraldehyde	−97	75	2.48	<− 5	1.4	12.5	42	380	230	T3	IIA
carbon disulphide	−112	46	2.64	<−20	1.0	60	30	1900	100	T5	*†
carbon monoxide	−205	−191	0.97	—	12.5	74.2	145	870	605	T1	IIB
chlordimethyl ether	—	—	—	—	—	—	—	—	—	—	IIA
chlorobenzene	−45	132	3.88	28	1.3	7.1	60	520	637	T1	IIA
chlorobutane	−123	78	3.2	<0	1.8	10.1	65	390	(460)	(T1)	IIA
chloroethane	−136	12	2.22	—	3.6	15.4	95	400	510	T1	IIA
chloroethanol	−70	129	2.78	55	5	16	160	540	425	T2	IIA
chloroethylene	−154	−14	2.15	—	3.8	29.3	95	770	470	T1	IIA
chloromethane	−98	−24	1.78	—	10.7	13.4	150	400	625	T1	IIA
chloropropane	−123	47	2.7	<−20	2.6	11.1	70	300	520	T1	IIA
coal tar naptha	—	—	—	—	—	—	—	—	272	T3	IIA

Table 10.1 *(continued)*

Compound	Melting point	Boiling point	Vapour density compared with air (air = 1)	Flash-point	Flammability limits in air				Ignition temp.	T class	Apparatus group
					Lower	Upper	Lower	Upper			
coke oven gas	–	–	–	–	–	–	–	–	–	–	*
cresol	– 11	191	3.73	81	1.1	–	45	–	555	T1	IIA
cyclobutane	– 91	13	1.93	–	1.8	–	42	–	–	–	IIA
cyclohexane	7	81	2.9	–18	1.2	7.8	40	290	259	T3	IIA
cyclohexanol	24	161	3.45	68	1.2	–	–	–	300	T2	IIA
cyclohexanone	– 31	156	3.38	43	1.4	9.4	53	380	419	T2	IIA
cyclohexene	–104	83	2.83	<–20	1.2	–	–	–	(310)	(T2)	IIA
cyclohexylamine	– 18	134	3.42	32	–	–	–	–	290	T3	IIA
cyclopropane	–127	– 33	1.45	–	2.4	10.4	40	185	495	T1	IIB
decahydronaphthalene	– 43	196	4.76	54	0.7	4.9	40	280	260	T3	IIA
diacetone alcohol	– 47	166	4.0	58	1.8	6.9	–	–	640	T1	IIA
diaminoethane	8	116	2.07	34	–	–	–	–	385	T2	IIA
diamyl ether	– 69	170	5.45	(57)	–	–	–	–	170	T4	IIA
dibutyl ether	– 95	141	4.48	25	1.5	7.6	48	460	185	T4	IIB
dichlorobenzene	– 18	179	5.07	66	2.2	9.2	130	750	(640)	(T1)	IIA
dichloroethane	– 98	57	3.42	–10	5.6	16	225	660	440	T2	IIA
dichloropropane	– 80	96	3.9	15	3.4	14.5	160	690	555	T1	IIA
diethylamine	<– 80	56	2.53	<–20	1.7	10.1	50	305	(310)	(T2)	IIA
diethylaminoethanol	–	161	4.04	(60)	–	–	–	–	–	–	IIA
diethyl ether	–116	34.5	2.55	<–20	1.7	36	50	1100	170	T4	IIB
diethyl oxalate	– 41	180	5.04	76	–	–	–	–	–	–	IIA
diethyl sulphate	– 25	208	5.31	104	–	–	–	–	–	–	IIA

*Apparatus group will depend on the relative proportions of the constituent gases.

Table 10.1 *(continued)*

Compound	Melting point	Boiling point	Vapour density compared with air (air = 1)	Flash-point	Flammability limits in air				Ignition temp.	T class	Apparatus group
					Lower	Upper	Lower	Upper			
dihexyl ether	−43	227	6.43	75	−	−	−	−	185	T4	IIA
di-isobutylene	−106	105	3.87	(2)	−	−	−	−	(305)	(T2)	IIA
dimethylamine	−92	7	1.55	−	2.8	14.4	52	270	(400)	(T2)	IIA
dimethylaniline	2	194	4.17	63	1.2	7	60	350	370	T2	IIA
dimethyl ether	−141	−25	1.59	−	3.7	27.0	38	520	−	−	IIB
dipropyl ether	−122	90	3.53	<21	−	−	−	−	−	−	IIB
dioxane	10	101	3.03	11	1.9	22.5	70	820	379	T2	IIB
dioxolane	−26	74	2.55	(2)	−	−	−	−	−	−	IIB
epoxypropane	−112	34	2.0	<−20	2.8	37	45	580	430	T2	IIB
ethane	−183	−89	1.04	−	3.0	15.5	37	195	515	T1	IIA
ethanol	−144	78	1.59	12	3.3	19	67	290	425	T2	IIA
ethanolamine	10	172	2.1	85	−	−	−	−	−	−	IIA
ethoxyethanol	−	135	3.1	95	1.8	15.7	75	−	235	T3	IIB
ethyl acetate	−83	77	3.04	−4	2.1	11.5	75	420	460	T1	IIA
ethyl acrylate	>−75	100	3.45	9	1.8	−	74	−	−	−	IIB
ethylbenzene	−95	136	3.66	15	1.0	6.7	44	−	431	T2	IIA
ethyldigol	−	202	4.62	94	−	−	−	−	−	−	IIA
ethylene	−169	−104	0.97	−	2.7	34	31	390	425	T2	IIB
ethylene oxide	−112	11	1.52	−	3.7	100	55	1820	440	T2	IIB
ethyl formate	−80	54	2.55	−20	2.7	16.5	80	500	440	T2	IIA
ethyl mercaptan	−148	35	2.11	<−20	2.8	18	70	460	295	T3	IIA

Table 10.1 *(continued)*

Compound	Melting point	Boiling point	Vapour density compared with air (air = 1)	Flash-point	Flammability limits in air				Ignition temp.	T class	Apparatus group
					Lower	Upper	Lower	Upper			
ethyl methyl ether	–	8	2.07	–	2.0	10.1	49	255	190	T4	IIB
ethyl methyl ketone	– 86	80	2.48	– 1	1.8	11.5	50	350	505	T1	IIA
formaldehyde	–117	– 19	1.03	–	7	73	87	910	424	T2	IIB
formdimethylamide	– 61	153	2.52	58	2.2	16	70	500	440	T2	IIA
hexane	– 95	69	2.79	–21	1.2	7.4	42	265	233	T3	IIA
hexanpl	– 45	157	3.5	63	1.2	–	–	–	–	–	IIA
heptane	– 91	98	3.46	– 4	1.1	6.7	46	280	215	T3	IIA
hydrogen	–259	–253	0.07	–	4.0	75.6	3.3	64	560	T1	IIC
hydrogen sulphide	– 86	– 60	1.19	–	4.3	45.5	60	650	270	T3	IIB
isopropylbitrate	–	105	–	20	2	100	–	–	175	T4	IIB
kerosene	–	150 to 300	–	38	0.7	5	–	–	210	T3	IIA
metaldehyde	246	112	6.07	36	–	–	–	–	–	–	IIA
methane (firedamp)	–182	–161	0.55	–	5	15	–	–	595	T1	I
methane (industrial)*	–	–	–	–	–	–	–	–	–	T1	IIA
methanol	– 98	65	1.11	11	6.7	36	73	350	455	T1	IIA
methoxyethanol	– 86	124	2.63	39	2.5	14	80	630	285	T3	IIB
methyl acetate	– 99	57	2.56	–10	3.1	16	95	500	475	T1	IIA
methyl acetoacetate	–	170	4.0	67	–	–	–	–	280	T3	IIA

*Industrial methane includes methane mixed with not more than 15% by volume of hydrogen.

Table 10.1 *(continued)*

Compound	Melting point	Boiling point	Vapour density compared with air (air = 1)	Flash-point	Flammability limits in air				Ignition temp.	T class	Apparatus group
					Lower	Upper	Lower	Upper			
methyl acrylate	<−75	80	3.0	−3	2.8	25	100	895	−	−	IIB
methylamine	−92	−6	1.07	−	5	20.7	60	270	430	T2	IIA
methylcyclohexane	−127	101	3.38	−4	1.15	6.7	45	−	260	T3	IIA
methylcyclohexanol	−38	168	3.93	68	−	−	−	−	295	T3	IIA
methyl formate	−100	32	2.07	<−20	5	23	120	570	450	T1	IIA
Naphtha	−	35 to 60	2.5	−6	0.9	6	−	−	290	T3	IIA
naphthalene	80	218	4.42	77	0.9	5.9	45	320	528	T1	IIA
nitrobenzene	6	211	4.25	88	1.8	−	90	−	480	T1	IIA
nitroethane	−90	115	2.58	27	−	−	−	−	410	T2	IIB
nitromethane	−29	101	2.11	36	−	−	−	−	415	T2	IIA
nitropropane	−108	131	3.06	(49)	−	−	−	−	420	T2	IIB
nonane	−54	151	4.43	30	0.8	5.6	37	300	205	T3	IIA
nonanol	−	178	4.97	75	0.8	6.1	−	−	−	−	IIA
octaldehyde	−	163	4.42	52	−	−	−	−	−	−	IIA
octanol	−16	195	4.5	81	−	−	−	−	−	−	IIA
paraformaldehyde	−	25	−	70	−	−	−	−	300	T2	IIB
paraldehyde	12	124	4.56	17	1.3	8.0	70	−	235	T3	IIA
pentane	−130	36	2.48	<−20	1.4	−	41	240	285	T3	IIA
pentanol	−78	138	3.04	34	1.2	10.5	44	380	300	T2	IIA
petroleum	−	−	−	<−20	−	−	−	−	−	−	IIA

Table 10.1 *(continued)*

Compound	Melting point	Boiling point	Vapour density compared with air (air = 1)	Flash-point	Flammability limits in air Lower	Upper	Lower	Upper	Ignition temp.	T class	Apparatus group
phenol	41	182	3.24	75	—	—	—	—	605	T1	IIA
propane	−188	−42	1.56	—	2.0	9.5	39	180	470	T1	IIA
propanol	−126	97	2.07	15	2.15	13.5	50	340	405	T2	IIA
propylamine	−101	32	2.04	<−20	2.0	10.4	49	260	(320)	(T2)	IIA
propylene	−185	−48	1.5	—	2.0	11.7	35	210	(455)	(T1)	IIA
propyl methyl ketone	−78	102	2.97	(16)	1.5	8.2	53	300	505	T1	IIA
pyridine	−42	115	2.73	17	1.8	12.0	56	350	550	T1	IIA
styrene	−31	145	3.6	30	1.1	8.0	45	350	490	T1	IIA
tetrahydrofuran	−108	64	2.49	−17	2.0	11.8	46	360	(260)	(T3)	IIB
tetrahydrofurfuryl alcohol	—	178	3.52	70	1.5	9.7	60	410	280	T3	IIB
toluene	−95	111	3.18	6	1.2	7	46	270	535	T1	IIA
toluidine	−16	200	3.7	85	—	—	—	—	480	T1	IIA
town gas (coal gas)*	—	—	—	—	—	—	—	—	—	—	IIB
triethylamine	−115	89	3.5	<0	1.2	8	50	340	—	—	IIA
trimethylamine	−117	3	2.04	—	2.0	11.6	49	285	(190)	(T4)	IIA
trimethylbenzene	−45	165	4.15	—	—	—	—	—	470	T1	IIA
trioxane	62	115	3.11	(45)	3.6	29	135	1100	410	T2	IIB
turpentine	—	149	—	35	0.8	—	—	—	254	T3	IIA
xylene	−25	144	3.66	30	1.0	6.7	44	335	464	T1	IIA

*Containing not more than 57% by volume of hydrogen and not more than 16% by volume of carbon monoxide, the remainder being a mixture of paraffin hydrocarbons and inert gas.

Table 10.2
EXTENT OF DAMAGE AND APPLIANCE OF ORIGIN

Appliance group	Damage					
	Confined to appliance only	Structure only	Contents only	Structure and contents	Burns to person only*	Total
Cooking	556	82	428	340	–	1406
Space heating	72	60	629	578	6	1345
Radio and television	429	12	118	175	–	734
Refrigerator	577	16	15	36	–	644
Motor	461	21	32	48	–	562
Light	260	68	115	75	–	518
Blanket	17	2	258	52	–	329
Iron	9	27	107	111	1	255
Generation and transmission	101	8	9	24	–	142
Leads and plugs	59	16	42	45	1	163
Water heating	61	10	31	48	–	150
Industrial	43	6	29	17	–	95
Driers	19	5	10	12	–	46
Hospital	9	–	2	2	–	13
Sundry	47	3	15	21	–	86
Unknown	–	–	–	1	–	1
Total No.	2720	336	1840	1585	8	6489
Per cent	41.9	5.2	28.4	24.4	0.1	100.1

*'Burns to person only' was not specified on the original self-coding questionnaire and may, therefore, have been included under another heading by some Brigades. (*Based on Fire Brigade Reports*)

Table 10.1 gives values for the minimum nominal spark ignition energy in air for a number of gases and vapours commonly used in industry. Table 10.2 shows an interesting typical 'breakdown' relating to electrical accidents.

10.4 Explosion and detonation

The simple picture outlined above must, however, be modified slightly if the process is to be properly understood. In the first place the energy depends to some extent on the physical properties and configuration of the apparatus used. For example, the igniting power of a long thin spark and a short fat spark are not precisely the same (Morgan and Taylor Jones[4,5]). Secondly combustion is not a simple process, the rearrangement of the atoms takes place one step at a time and some steps are reversible, according to the law of mass action, so that a whole series of transient intermediate products are formed. For example, when hydrogen burns to form water vapour, superficially a very simple chemical

reaction, the following detailed reactions may, according to circumstances, be going on simultaneously (Bradley[6]):

$$*H + O_2 \rightarrow OH + H$$
$$O + OH \rightarrow O_2 + H$$
$$*O + H_2 \rightarrow OH + H$$
$$H + OH \rightarrow H_2 + O$$
$$OH + H_2 \rightarrow H_2O + O$$
$$H + H_2O \rightarrow H_2 + OH$$
$$OH + OH \rightarrow H_2O + O$$
$$O + H_2O \rightarrow OH + OH$$

In some reactions a single active particle (e.g. a radicle with an unsatisfied valency bond) is formed to carry on the chain. In others two such radicles are formed to initiate a pair of branching chains. On the other hand a chain reaction may come to an end because a reactive radicle is removed, or activation energy is absorbed. This may result from a chemical reaction or by contact with a solid surface — such as the wall of an enclosure, which acts as an anti-catylyst. This has a direct bearing on the design of flameproof enclosures.

A thermal explosion occurs when no stable balance is reached between the rate of generation of heat and the loss by radiation, conduction, etc and the temperature rises until it is halted by the availability of fuel and/or oxygen. A branching chain explosion occurs when the rate of initiation of new chains exceeds the rate at which chains are ended. These are characterised by a rapid rise of pressure and a high flame velocity. This is very apparent with hydrogen/oxygen and hydrogen/air mixtures and such explosions are very destructive, and may develop into shock and detonation waves. In many explosions both these processes probably occur.

It is well-known that when a column of air, such as a lightning channel, is suddenly raised to a very high temperature it expands extremely rapidly and sends out a shock wave which has a very steep wave front and travels at several times the velocity of sound; but as it travels out it loses energy and subsides into an ordinary sound wave train. If, however, a shock wave passes through an inflammable gas/air mixture, the mixture will be compressed and raised far above its ignition temperature and burn explosively, liberating further energy, maintaining or increasing the temperature, pressure and velocity; this is a detonation. Where a gas, such as hydrogen or oxygen, is ignited the speed of burning may be sufficient to initiate a shock wave and ultimately detonation. The same mechanism occurs when a high-explosive solid such as gelegnite 'burns to detonation'.

As a result of the very rapid rise of pressure a high velocity hydrogen explosion or detonation has a far more shattering effect than a normal gas explosion.

10.5 Delay of ignition

A flame or explosion does not immediately follow a spark, or the application of other sources of heat, owing to the reaction time. This may be very short or quite considerable, depending on the amount and rate of supply of energy and other considerations. Although this is not often an important matter, it may affect the probability of ignition in some marginal cases, e.g. with a rapidly moving friction spark, or flow of gas across a hot wire.

10.6 Limits of flammability [7]

The ignition temperature, flame velocity and pressure developed depend on the proportion of gas to air and there are limits outside which no ignition is possible. The most easily ignited mixture is not usually the same as the theoretical mixture, i.e. that providing just the correct amount of oxygen for complete oxidation. The limits of flammability of a wide range of industrial gases and vapours are given in Table 10.1. It will be noted that the range for hydrogen is very wide, 4 to 76% by volume, whereas for methane it is 5 to 15%. This is very important since it means that almost any mixture of hydrogen and air can be ignited whereas this is not possible with most mixtures of methane and air.

Town gas, which has a high proportion of hydrogen, about 60%, was much easier to ignite than North Sea gas which is primarily methane. This and relatively low flame velocity has a direct bearing on the use of flame failure devices in furnaces.

10.6.1 Calculating flammability limits [27]

It is useful to have a method of calculating the lower flammability limit of particular flammable gases and vapours if tables of data are not available or the particular substance is not included in available tables.

Many mathematical equations have been prepared to enable the limit to be calculated from a knowledge of the chemical formula of the substance. Unfortunately, no simple equation exists which can be applied

to all flammable gases and vapours and will give accurate results in all cases.

One of the simpler guides is the following: a vapour/air mixture will fail to ignite if the concentration of the vapour is reduced to about half that required for complete combustion, otherwise known as the stoichiometric concentration. More precisely, if N is the number of oxygen atoms required for the complete combustion of 1 molecule of vapour then:

$$\text{Lower limit concentration} = \frac{100}{4.85N-1.425}\%$$

Taking methane as an example, we require to determine N.
The chemical equation for the complete combustion of methane is given by:

$$CH_4 + 2O_2 = CO_2 + 2H_2O$$

i.e., each molecule of methane requires 2 molecules or 4 atoms of oxygen for complete combustion. In this case, therefore, $N = 4$.
If we now substitute $N = 4$ in the equation, we obtain a value for the lower limit concentration of 5.56%.

This calculated result can be compared with published experimental data which gives 5.3% as a typical value. Similar calculations for other flammable vapours and gases can be carried out and Table 10.3 compares some of these with lower flammable limits determined by actual tests.

Table 10.3
COMPARISON OF FLAMMABLE VAPOURS AND GASES

Substance	N	Lower flammable limit Calculated	Lower flammable limit Experimental
Methane CH_4	4	5.56	5.3
Ethane C_2H_6	7	3.08	3.0
Propane C_3H_8	10	2.12	2.2
Butane C_4H_{10}	13	1.62	1.8
Pentane C_5H_{12}	16	1.31	1.5
Ethylene C_2H_4	6	3.61	3.1
Propylene C_3H_6	9	2.37	2.4
Butylene C_4H_8	12	1.76	1.6 (α) 1.8 (β)
Acetylene C_2H_2	5	4.38	2.5
Methanol CH_3OH	3	7.62	7.3
Ethanol C_2H_5OH	6	3.61	3.3
n-propanol C_3H_7OH	9	2.37	2.1
Benzene C_6H_6	15	1.40	1.4
Toluene $C_6H_5CH_3$	18	1.16	1.3
Diethyl ether $C_2H_5OC_2H_5$	12	1.76	1.85

Although the formula is very simple, it gives fairly good results with a large number of paraffin hydrocarbons, aromatic hydrocarbons, alcohols and other substances. However, it is not universally applicable as shown by the rather large discrepancy between the calculated and actual results for acetylene.

10.6.2 Extinction of gases and vapours

A flame will be extinguished either if the supply of fuel is cut off, if it is diluted until it is no longer combustible, or if it is cooled. Figures 10.1(a) and (b) shows the effects of dilution. Dilution with CO_2 and H_2O are of particular interest because they can be released as a fine spray in the case of water and as 'snow' in the case of CO_2. These have been successfully used in suppressing oil fires in electricity substations.

Where it is necessary to prevent ignition by using an inert atmosphere, nitrogen is often the first choice. The easiest precaution may be to use air from which the oxygen has been 'burned out', but care must be taken in the choice of fuel since the products of combustion may be corrosive or otherwise objectionable.

10.6.3 Gas mixtures[8]

Some gases even in very small amounts, may have an influence on ignitability, presumably because they assist or inhibit the production of branching chain reactions. For example, 5 % of H_2O increases the ignition temperature of hydrogen by about 10°C but it lowers the ignition temperature of CO at atmospheric pressure. NO and NO_2 have similar effects, but these are not usually sufficiently great to have much influence on our problems. (The effect of tetra ethyl lead on combustion in an engine cylinder may however be noted.)

Otherwise the effects on the limits of inflammability of mixing gases or vapours is fairly simple in the case of some vapours, such as the paraffin series for which

$$L = \frac{100}{P_1/N_1 + P_2/N_2 + P_3/N_3 \text{ etc.}}$$ (Chatillier's Law) (10.1)

where
L = inflammable limit for mixture;
P_1, P_2, etc, are the proportions of the constituents and
N_1, N_2, etc are respective limits of inflammability.
This empirical relation works quite well in many cases but is not universally true (see Coward and Jones[7]). Sometimes, however, the error is very great as for H_2 with H_2S.

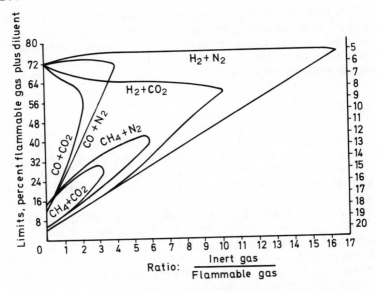

Figure 10.1(a) Limits of hydrogen, carbon monoxide and methane containing various amounts of carbon dioxide and nitrogen

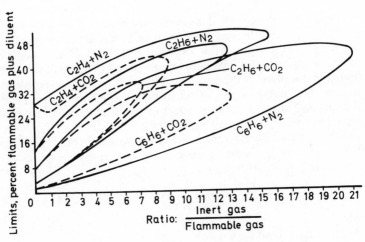

Figure 10.1(b) Limits of flammability of ethane, ethylene and benzene containing various amounts of carbon dioxide and nitrogen

(Figure 10.1(a) and (b) from BS 5345 Pt 1 1976 are reproduced by permission of the British Standards Institute, 2 Park Street, London W1A 2BS, from whom complete copies can be obtained)

The limits of inflammability give a useful, if not infallible, guide to probable difficulties met in designing flameproof enclosures. Hydrogen, with wide limits, is very difficult, methane with narrow limits is easy. A mixture of 60 % hydrogen and 40 % methane which is approximately the constitution of 'coal gas', is fairly difficult but not impossible whereas blue water gas, which is a mixture of hydrogen and CO is nearly as difficult as pure hydrogen although CO has a much lower calorific value than methane; its limits of inflammability are, however, relatively wide i.e. 12.5–74%. This is no doubt a result of a branching chain reaction.

10.7 Flash point

The significance of the flash point of a liquid (see Table 10.1) is often misunderstood. Flash point is not the ignition temperature but the temperature at which the vapours over the free surface of the liquid can be ignited by a flame in a standardised apparatus. It is not, therefore, a fundamental property of the liquid but it is approximately the temperature at which the vapour pressure produces (in still air) a concentration sufficiently above the lower inflammable limit for it to be effectively ignited. It is a useful, but not exact, indication of the lowest temperature at which the liquid is easily ignited, but it must be remembered that in many industrial situations the temperature of liquids and their immediate vicinity is *above* the ambient temperature of the surroundings. The flash point is considerably lower than both the ignition temperature and the boiling point (i.e. the temperature at which the vapour pressure equals the atmospheric pressure).

In practice a liquid is not effectively ignited unless the flame caused by igniting the free vapour boils off sufficient extra vapour to make it stable; this depends on air movements. The following figures illustrate the relationship:

Ethyl Alcohol

Auto ignition temperature app.	450°C
Boiling point	78°C
'Closed' flash point	12°C

10.8 Corona, glow discharges and silent discharges

Sloane[8] demonstrated that both transient and steady state glow discharges would ignite a coal gas/air mixture. This would occur when the discharge was between very poor conductors such as slate pencils

(though not between perfect insulators). I have obtained streams of sparks 11 000 V between glass rods well below red heat and even from the vitreous glaze of chips of porcelain. Glass is a poor electrolytic conductor but improves greatly at raised temperatures. Sutcliffe[9] ignited hydrogen and, I believe coal gas, by a silent (invisible) discharge only otherwise detectable as an electric wind; Klinkenberg[10] ignited butane/air mixture. Sloane and Klinkenberg used short gaps between electrodes whereas Sutcliffe used quite considerable distances; thus getting conditions closer to overhead line conductors or St Elmo's fire.

From his experiments Sloane concluded that the ignition energy was 0.003 joules for intermittent discharges between slate pencils, but allowing for I^2R losses the true figure must have been substantially lower. This is about an order of magnitude higher than the minimum for spark ignition, but being more diffuse it would have been less efficient and it is probable that the most favourable conditions were not obtained. Sloane obtained ignition with from 150 to 300 micro-amp steady discharges; Klinkenberg's figures are about a tenth of this.

The above are high voltage effects in rather unusual conditions, but it is clear that, where high voltage conductors or insulators are exposed to explosive atmospheres, there is a definite hazard. A case in point is the electrical connections of a hydrogen-cooled alternator if there is a gas leakage.

10.9 Cool flames and peroxides

This is a rather complex subject. It has been found that some organic vapours will burn at a temperature well below their normal ignition temperature. For the most part, this occurs with over-rich mixtures and, in the case of most vapours, does not occur at atmospheric pressure; there are, however, a few exceptions, notably acetaldehyde, diethyl ether and divinyl ether which have cool flame zones at normal pressures. Such flames progress slowly and give out very little light so that they are not easily noticeable, but if they reach a place where the concentration is suitable they may change into normal flame.

On the other hand, if an over-rich mixture is compressed, heated to produce a cool flame and then released into free air, forming a weaker mixture, a normal flame or explosion may occur. This can sometimes be the explanation of insulating oil explosions since members of the paraffin series have cool flame zones at raised pressures. (This is discussed further in section 10.15).

An alternative explanation of low temperature ignition of oil vapour may be the presence of NO^2 acting as a catalyst, or peroxides. Ethyl ether peroxide, for example, may explode spontaneously, and if present

in small quantities may cause the ignition of ether/air mixtures at low ambient temperatures.

10.10 Hot surfaces

Gases and vapours may oxidise catalytically on hot surfaces below their normal ignition temperature and outside the normal limits of flammability. This causes a further rise in temperature of a surface such as a heated platinum wire, and the effect is used as a measure of the proportion of flammable vapour in air in a number of industrial instruments.

In free air, if the rise in temperature is sufficient, the hot surface may cause a self-propagating flame or explosion. The temperature of a hot wire necessary to produce ignition does not correspond accurately with the normal ignition temperature because the process is complex. For example, the initiation of a flame depends on the flow of air and vapour mixture over the wire, the diffusion of fuel to the surface and of heat and products of combustion away from it.

A hot wire (e.g. a radiator element) needs to be initially at well above 1000°F to ignite methane. In some rather rough-and-ready experiments I found that an electric radiator element which would not ignite methane would ignite petrol vapour if it was bright but not if it was dull and it ignited the vapour from some heavier liquids quite easily. Hydrogen and coal gas ignite quite easily in this way as does ethyl ether.

I next took a piece of cottonwool, dipped it in a flammable liquid and held it near a radiator element, it often would not ignite if too close, no doubt because the vapour formed was over-rich, but it did ignite if held at a distance of 5 or 10 mm. For heavy liquids such as olive oil or turpentine, there was a notable delay while they warmed up. Possibly if the experiment had been conducted in a different way the results could also have been different; it is a matter of the conditions being just right. This variability explains the lack of consistency found when investigating fires.

The importance of this matter is illustrated by a number of accidents. On one occasion when a quart of ether was upset in a hospital, the heavy vapour flowed a considerable distance along the floor and was ignited by an electric radiator. A nurse was severely burnt. On another occasion when a similar container of inflammable liquid was upset in a small laboratory, a chemist was fatally burned.

A foreman upset some petrol on his overalls. Soon afterwards he went into the office and stood in front of a radiator. His overalls caught fire and he was fatally burned. Had the radiator elements been a little cooler (e.g. as a result of voltage reduction) there might well have been no fire.

Another aspect is the possibility of igniting gas/air mixtures by the hot filament of a broken electric lamp bulb. Alsop found that the bulb remained hot enough to ignite methane for about 1/5th second after the glass envelope was broken. This has an important bearing on lighting areas where full flameproof enclosure is not used (see Chapter 11).

The possibility of igniting such materials as carbon disulphide, ether, or ethyl nitrite by hot surfaces has recently been considered in the classification of 'flameproof' equipment and 'increased safety' (see Chapter 11).

10.11 Friction sparks

Just 200 years ago Volta demonstrated that marsh gas and hydrogen could be ignited by sparks produced by the impact of steel on stone and drew attention to the possible danger of lighting colliery workings by 'steel mills' which produced a continuous shower of sparks, this being then considered to be safer than a flame. An explosion at Wallsend colliery in 1785 was attributed to this cause and led to the invention of the safety lamp by Humphrey Davy (and simultaneously by George Stephenson).

Although this is not strictly an electrical matter, mechanical sparks may be caused by electrical apparatus and it is sometimes necessary to distinguish between causes. A mechanical spark may occur, for example, if the rotor bars of an induction motor are displaced and touch the stator or by a collapsed bearing.

Another important example has been the possibility of causing a spark by dropping a 'safe' metal encased portable lamp or torch in a hazardous situation. A considerable amount of research has been carried out on these matters at the Safety in Mines Research Establishments in Sheffield and Buxton. A series of their reports and the conclusions were summarised by Titman in a paper to the Institution of Mining Engineers[11]. The important point here is that light metals and alloys based on magnesium, aluminium and titanium when struck obliquely against rusted steel give sparking which is a result both of a thermite reaction and burning metal which the heat of this reaction ignites. A similar result is obtained by striking slightly rusty steel on which there is a 'smear' of light metal or alloy or aluminium paint, but not on chemically clean steel. It has also been found that a thermite type reaction may occur when siliceous materials are struck with a light metal or alloy, the oxygen being obtained by reduction of silica. The lesson is that appliances which may be dropped should be made from other materials or covered in plastic. All-insulated electrical appliances seem to be particularly suitable.

To complete the picture, moving hot iron particles (sparks) do not ignite methane easily, presumably because of the delay in ignition; i.e. the particles have passed on before the surrounding gas can be heated up and fired. But one must remember the tinder box effect. Dr Coward of the Safety in Mines Research Board in conversation told me of a demonstration when a steel wheel in contact with a rail was rapidly revolved producing showers of sparks without effect, but when they were 'collected' on a newspaper or rag ignition followed.

10.12 Ignition of wood, varnishes, textiles, etc[12]

When soft woods such as pine are heated, inflammable vapour is distilled off which consists largely of methyl alcohol with some acetone, together with acetic acid, water vapour etc. This commences at about 250°C, and below 300°C its evolution is 'explosive'. The vapour is easily ignited. Solid seasoned hard woods, on the other hand, may char on the surface but burn slowly. A 1½ in oak door is a good fibre barrier and an oak beam may still carry a load after a rolled steel joist has buckled and collapsed. The wooden door is in fact a better fire barrier than a steel plate because the latter may become red hot.

Where wood or other materials are covered with varnish or lacquer, particularly fresh shellac varnish or cellulose paints and lacquers, a very dangerous condition can arise as these substances may break down and give off inflammable vapours at quite low temperatures. Celluloid is not stable above 80° to 100°C. The Fire Protection Association[13] reporting on experimental fires state:

'After being confined for about 15 minutes to the corner group of furniture in which it originated, the fire suddenly spread, in what has been called a 'flash over' involving the whole contents of the room at once . . . The flash spread . . . was due to preheating by the primary fire . . . which gradually raised the temperature of the remaining woodwork to ignition point. Some of the furniture reached a rather higher temperature and began to distil without actually igniting . . . For this flash over to occur, there must be an adequate supply of oxygen in the room. The oxygen content of air is about 21 % by volume and if it drops to 15% flame combustion can no longer be supported. Thus in a room of 2000 cubic feet capacity . . . the oxygen . . . would barely suffice to burn 7 lb of wood. When the door is opened extra oxygen – or perhaps air movement will often cause flash over. This phenomenon, not previously explained, is well-known to firemen.'

An industrial example of flash over occurred in a railway coach. There had apparently been an unnoticed smouldering fire underneath a seat and sufficient vapour had been distilled off from the (presumably cellulose) varnish to cause a flash over when a workman entered and

attempted to use an electric drill or paint stripper. Similar trouble has occurred with shellac varnish and impregnating coil insulation of electric apparatus. For example, following an explosion in a gas-tight switchbox in which it was believed that tracking on the surface of the insulating material caused flammable vapour to be evolved. Tests showed that hydrogen and carbon monoxide could be evolved together with considerably smaller quantities of methane, acetylene and other hydrocarbons. I have myself investigated an explosion apparently caused by the boiling off of solvent from an imperfectly baked armature coil impregnated with shellac varnish. The use of shellac varnish has been largely superseded for insulation, but old equipment still exists.

Attention may be drawn here to the flammability of some synthetic materials used for upholstery, carpets, curtains and wall coverings. A rather strange example was a fire in an interior-sprung mattress caused by the use of high frequency physiotherapy apparatus which induced sparks among the springs. There was no visible fire on the outside but when opened up the fabric enclosing the springs was glowing.

Rubber insulation when heated out of contact with air gives off fumes which may ignite spontaneously when they come into contact with air. (See also under Cables, section 13.3).

10.13 Spontaneous combustion at low temperatures

Some materials catch fire at comparatively low ambient temperatures. Among gases and vapours, carbon disulphide has an ignition temperature of approx. $100°C$ and could be ignited by an unlagged steam pipe; for ethyl nitrite the temperature is $32°C$, i.e. below the boiling point of water. It is shown in section 10.15 that the vapour from overheated insulating oil has ignited at $145°C$. The presence of traces of peroxides, NO_2 etc. may substantially lower auto-ignition temperatures and the presence of ether oxide in ethyl ether is stated to have a similar effect.

With solid substances, danger chiefly arises from granular material or dusts with a large surface area per unit weight. Many organic materials will oxidise slowly but continuously in air. In dust deposits, the temperature will rise if there is sufficient depth to conserve heat with sufficient porosity to allow an adequate supply of oxygen and the removal of the products of combustion (CO_2, CO, H_2O etc.). Thus the temperature will rise until equilibrium is reached with generation and loss of heat in balance; if no balance is reached the material will eventually catch fire. Because reaction rates double for approximately every $10°C$ rise in temperature, this is particularly likely to occur when dust or granules rest on a heated surface. This has an important bearing in the initiation of dust explosions (section 10.14).

A typical and important example of this is coal dust. Experience shows that the reaction rate becomes important when the dust rests on a surface at about 80°C. This is only a rough approximation to give a sense of proportion and is not to be taken as 'design data' or a reliable criterion of safety.

Celluloid is liable to break down and may ignite spontaneously at very low temperatures. This applies to the old standard cinematograph film (though not to non-flam film) and also cellulose lacquer, particularly cellulose residues in paint spraying cubicles and departments. For this reason all surfaces which may come into contact with these materials must be kept well below 100°C (80°C has sometimes been used as the limiting value).

The ignition of textiles by radiant heat and conduction is important and it should be noted that if material is draped over convection heaters, restricting the transfer of heat, the temperature may rise until material scorches and eventually catches fire unless there is automatic temperature control.

Except for gold and platinum, which do not oxidise, metals very rapidly form a surface refractory oxide film. This film is often invisible but even so it will usually prevent or at least seriously retard further oxidation. Therefore the risk of spontaneous combustion does not normally arise, except in the cases of sodium and potassium, which are liable to catch fire spontaneously. For this reason these metals are often stored immersed in oil.

Magnesium, aluminium, and magnesium alloys if heated in a flame or by an electric arc, burn violently and since they decompose water when hot this must never be used to extinguish them, because of the probability of a hydrogen/air explosion. Their dusts are a serious explosion hazard and the course of ignition is not always clear. It has been suggested that if ground to powder in an inert atmosphere or with restricted air access, the new surfaces may not be able to form a protective oxide film. If the powder is then suddenly dispersed in air the rapid surface oxidation may raise the particles to ignition temperature because of the high ratio of surface to bulk. This has not been proved, I believe, but zirconium powder does 'explode' spontaneously if 'puffed' into free air.

Titanium is chemically related to zirconium and can also be a fire risk. This is also true of the higher alkali metals rubidium and caesium; the reaction possibly involves a reaction with atmospheric moisture.

10.14 Dust and droplet clouds and explosions

Dust cloud explosions are one of the chief causes of dangerous explosions in coal mines. In industry they are also a major hazard and are

Table 10.4a
SUMMARY OF FLAMMABILITY AND EXPLOSIBILITY OF POWDERS USED IN THE PLASTICS INDUSTRY
(From United States Bureau of Mines Report of Investigations 3751)

Type of powder*	Relative flammability‡ (per cent inert)		Ignition temperature¢ of dust cloud		Minimum energy required for ignition‖	Minimum explosive concentration‖	Maximum pressure†	Rate of pressure rise† lb/in²/sec	
	In furnace	In spark apparatus	°C	°F	Joule	oz/ft³	lb/in²	Average	Maximum
Resins									
Shellac, rosin, gum	90+	90+	390	735	0.01	0.015	58	1 240	2 990
Phenolic	90+	90	500	930	0.01	0.025	61	1 370	3 160
Coumarone-indene	90+	90+	520	970	0.01	0.015	63	1 370	2 990
Cellulose acetate	90+	83	410	770	0.015	0.035	68	800	1 740
Lignin	90+	78	450	840	0.02	0.040	69	760	2 700
Vinyl butyral	90+	80	390	735	0.01	0.020	60	470	1 020
Polystyrene	90+	90+	490	915	0.12	0.020	44	350	650
Urea	85	60	470	880	0.08	0.070	65	340	850
Vinyl	90+	73	550	1 020	0.16	0.040	49	250	490
Chlorinated paraffin	0	0	840	1 545	–	–	0	0	0
Moulding compounds									
Cellulose compounds	90+	90	320	610	0.01	0.025	62	1 180	2 260
Phenolic	90+	80	490	915	0.01	0.030	63	900	2 080
Synthetic rubber	90+	80	320	610	0.03	0.030	59	740	1 870

*The powders in each group are arranged approximately in the order of decreasing dust-explosion hazard.
†Pressure data obtained at a dust-cloud concentration of 0.5 oz/ft³.
‡Relative inflammability = % of added inert dust (fuller's earth) necessary to inhibit effective ignition.
¢Temperature of wall of small experimental furnace or muffle.
‖Energy of spark in condenser discharge.

Table 10.4a (continued)

Type of powder*	Relative flammability‡ (per cent inert)		Ignition temperature§ of dust cloud		Minimum energy required for ignition‖	Minimum explosive concentration	Maximum pressure†	Rate of pressure rise† lb/in²/sec	
	In furnace	In spark apparatus						Average	Maximum
Methyl methacrylate	90+	78	440	825	0.015	0.020	57	570	1 200
Polystyrene	90+	90+	560	1 040	0.04	0.015	50	740	1 640
Urea	90+	68	450	840	0.08	0.075	63	710	1 800
Vinyl	5	0	690	1·275	–	–	0	0	0
Primary ingredients of resins									
Hexamethylenetetramine	90+	90+	410	770	0.01	0.015	64	940	2 570
Pentaerythritol	90+	90	450	840	0.01	0.030	65	980	2 170
Phthalic anhydride	90+	90	650	1 200	0.015	0.015	49	1 270	1 690
Rennet casein	90	35	520	970	0.06	0.045	49	190	500
Fillers for moulding compounds									
Ground wood flour	90+	80	430	805	0.02	0.040	62	830	2 080
Ground cotton flock	90+	73	470	880	0.025	0.050	67	870	2 990
Ground alpha pulp	90+	85	480	895	0.08	0.060	60	520	1 450
Asbestos, asbestine, mica	No ignitions obtained in any test; these powders present no dust-explosion hazard.								

Table 10.4b
SUMMARY OF RELATIVE FLAMMABILITY AND EXPLOSIBILITY OF METAL POWDERS
(United States Bureau of Mines Report of Investigations 3722)

Metal	Relative flammability (per cent inert)		Ignition temperature of dust cloud		Maximum pressure	Rate of pressure rise, lb/in²/sec	
	In furnace	In spark tube	°C	°F	lb/in²	Average	Maximum
Zirconium, milled	90+	90+	Room*	Room*	42	600	2 570
Magnesium, stamped	90+	90+	520	970	72	1 450	4 760†
Magnesium, milled	90+	90	540	1 005	65	1 600	3 160
Dowmetal, milled	90	90	430	805	56	1 600	2 570
Magnesium-aluminium, milled	80	78	535	995	62	2 350	3 000
Aluminium, stamped	60	80	645	1 195	62	2 170	5 700†
Aluminium, atomized	10	78	700	1 290	58	1 050	2 250
Titanium, milled	55	78	480	895	52	750	1 640
Iron, reduced	90+	55	315	600	36	230	430
Iron, carbonyl	85	60	320	610	31	660	1 200
Manganese, milled	40	25	450	840	31	220	400
Zinc	35	45	600	1 110	36	710	1 350
Silicon, milled	0	33	775	1 425	62	440	1 180
Tin, atomized	10	30	630	1 165	34	500	850
Antimony, milled	65	18	415	780	20	90	150
Lead, stamped	20	0	580	1 075	‡	—	—
Lead, atomized	0	0	710	1 310	‡	—	—
Cadmium, atomized	18	0	570	1 060	‡	—	—
Copper, reduced	3	0	700	1 290	‡	—	—
Iron, milled	0	0	780	1 435	‡	—	—
Chromium, milled	0	0	900	1 650	‡	—	—

*Ignition may have resulted from static electricity in dust cloud.
†This extremely high value should be accepted with caution because of possible errors in determining maximum slope of pressure/time curves.
‡This powder not ignited by standard spark in Hartmann apparatus.

very destructive although not many industrial explosions have been *proved* to have been caused by electrical installations.. The potential hazard is, however, great and the risk must be taken seriously. A great deal of research work has been carried out (particularly on behalf of the Factory Inspectorate at the Safety in Mines Research Establishment) and was reviewed in 1962 by Brown and James[14].

Although the power behind a dust explosion may be very great the velocity of propagations is often relatively low. In one explosion in a flour mill, which I investigated, a foreman saw the advancing flame and dived to safety down a sack chute. The roof was, however, lifted bodily and dropped down several degrees out of line. On the other hand, metal dust explosions are often very violent. I visited one works where an aluminium powder sieving or grinding machine was segregated in an enclosure with 350 mm (14 in) reinforced walls. Nothing, it seemed, was left of the building much larger than a 7 lb biscuit tin (approximately 250 mm cube).

In a long list of fires and explosions in 'Fires Involving Dusts' issued by the Fire Protection Association[15], only one explosion was clearly attributable to electricity where 'discharge of static electricity ignited a filter bag used for collecting dust'. This instance is relevant to the section on 'static electricity' below. Significantly another was possibly a thermic reaction (see above) when a spark from a bricklayer's trowel ignited magnesium dust on a stone floor. Tables 10.4a and 10.4b give a number of materials which are known to produce dangerous dust clouds. (More recent research, with improved instruments, suggests that the instantaneous pressures may be substantially higher than those given in the tables.) For a general discussion of dust hazards, Safety Health and Welfare pamphlet 22 (New Series), *Dust explosions in factories*[16] and K.N. Palmer's book *Dust explosions and fire*[17] may be consulted. Suitable electrical equipment is dealt with below.

Tables 10.5 and 10.6 give further information on explosive dusts, from a different source. It will be noted that they do not agree exactly with Tables 10.3 and 10.4; this is usual in this field where materials differ both in composition and particle size. Very conveniently a concentration of one gram per litre is very close to one oz per cubic foot, which makes conversion between units simple. Table 10.6 gives some data for explosive vapours expressed in this way instead of as volume per cent. It will be seen that for organic materials the values are, with one or two exceptions, of the same order of magnitude for vapours and gases, which is to be expected as they have to combine with the same oxygen concentration.

In practice an explosive dust cloud consisting of, say, 0.05 oz per cubic foot would be comparable to a dense fog and cannot exist in normal working conditions, except perhaps locally or inside closed

Table 10.5
EXPLOSION CHARACTERISTICS OF VARIOUS DUSTS

Type of dust	Ignition temperature of dust cloud, degree C.	Minimum spark energy required for ignition of dust cloud, millijoules	Minimum explosives concentration, oz. per cu. ft.	Maximum explosion pressure, p.s.i.	Rates of pressure rise, p.s.i. per second		Limiting oxygen % to prevent ignition of dust cloud by electric sparks
					Average	Maximum	
METAL POWDERS							
Aluminium, atomized	700	50	0.040	58	1050	2250	3
Aluminium, stamped	645	20	0.035	89	2150	5700	
Iron, hydrogen reduced	315	160	0.120	29	650	1200	13
Magnesium, atomized	600	240	0.030	57	750	1450	3
Magnesium, milled	520	80	0.020	65	1500	3150	(b)
Magnesium, stamped	520	20	0.020	72	4400	4750	(b)
Manganese	450	120	0.210	25	200	300	15
Silicon	775	900	0.160	62	450	1200	15
Tin	630	160	0.190	26	250	400	16
Titanium	480	120	0.045	44	750	1100	(b)
Vanadium	500	60	0.220	34	200	300	13
Zinc	680	900	0.500	13	150	300	10
Zirconium	(a)	15	0.040	42	1450	4000	(b)
Dowmetal	430	80	0.020	56	1600	2550	(b)

(a) When zirconium powder was dispersed into air at room temperature, it ignited under some conditions, apparently due to static electric discharge between the particles in the dust cloud.

(b) The oxygen reduction tests were made in air-CO_2 mixtures. Dust clouds of zirconium, magnesium, titanium, and certain magnesium-aluminium alloys ignited in pure CO_2.

(Occupational Health Review.) 1959, **10** (4), 15.

Table 10.5 (continued)

Type of dust	Ignition temperature of dust cloud, degree C.	Minimum spark energy required for ignition of dust cloud, millijoules	Minimum explosives concentration, oz. per cu. ft.	Maximum explosion pressure, p.s.i.	Rates of pressure rise, p.s.i. per second		Limiting oxygen % to prevent ignition of dust cloud by electric sparks
					Average	Maximum	
Ferrotitanium, low-carbon	370	80	0.140	34	600	1400	13
Ferrosilicon (88% S)	860	400	0.425	36	200	300	19
Magnesium-aluminium (50-50)	535	80	0.050	61	2250	3000	(b)
PLASTICS							
Allyl alcohol resin	500	20	0.035	68	1750	3550	17
Casein	520	60	0.045	49	200	500	
Cellulose acetates	320	10	0.025	82	1200	2400	14.5
Cellulose propionate	460	60	0.025	66	1350	2350	
Coumarone-indene resin	520	10	0.015	63	1350	3000	14.5
Hexamethylenetetramine	410	10	0.015	64	950	2550	14.5
Lignin resin	450	20	0.040	69	750	2700	17
Methyl methacrylate	440	15	0.020	57	550	1200	14.5
Pentaerythritol	450	10	0.030	65	1000	2150	14.5
Phenolic resin	460	10	0.025	61	1350	3150	14.5
Phthalic anhydride	650	15	0.015	49	1250	1700	14.5
Polyethylene resin	450	80	0.025	83	400	1250	15
Polystyrene	490	120	0.020	44	350	650	17
Shellac, rosin, gum	390	10	0.015	58	1250	3000	14.5
Synthetic rubber, hard	320	30	0.030	59	750	1850	15
Urea moulding compound	450	80	0.075	63	700	1800	17
Vinyl butyral resin	390	10	0.020	60	450	1000	14.5

Table 10.5 (continued)

Type of dust	Ignition temperature of dust cloud, degree C.	Minimum spark energy required for ignition of dust cloud, millijoules	Minimum explosives concentration, oz. per cu. ft.	Maximum explosion pressure, p.s.i.	Rates of pressure rise, p.s.i. per second		Limiting oxygen % to prevent ignition of dust cloud by electric sparks
					Average	Maximum	
MISCELLANEOUS							
Aluminium stearate	400	15	0.015	62	750	2100	16
Coal, bituminous	610	40	0.035	46	350	800	15
Coal-tar pitch	–	80	0.080	49	350	650	15
Crude rubber, hard	350	50	0.025	57	850	3350	15
Dinitro-*ortho*-cresol	440	–	0.025	55	1300	2250	
Gilsonite	580	25	0.020	56	900	1850	
Phenothiazine	540	–	0.015	43	600	1450	16
Soap	430	60	0.045	60	660	1300	
Sulphur	190	15	0.035	41	700	1950	11

Table 10.6
MAXIMUM PERMISSIBLE OXYGEN CONTENT TO PREVENT THE IGNITION OF COMBUSTIBLE DUSTS

Type of dust	Using carbon dioxide for inerting				Using nitrogen for inverting	
	Hot surface (1560° F) ignition		Spark ignition		Spark ignition	
	O_2 percentage above which ignition can take place	Maximum recommended O_2 per cent	O_2 percentage above which ignition can take place	Maximum recommended O_2 per cent	O_2 percentage above which ignition can take place	Maximum recommended O_2 per cent
METAL POWDERS						
Aluminium (atomised)			3	2½	9	7
Magnesium			3	2	2	1½
Magnesium/aluminium alloy			0	0	6	5
Titanium			0	0	6	5
Zinc			10	8	10	8
Zirconium			0	0	4	3
RESINS						
Cellulose acetate	5	4	13	10½		
Phenolic	9	7	14½	11½		
Polystyrene	7	5½	14½	11½		
Urea	11	9	17	13½		
MOULDING COMPOSITIONS						
Cellulose acetate	7	5½	11½	9		
Methyl methacrylate	7	5½	14½	11½		
Phenolic	7	5½	14½	11½		
Polystyrene	9	7	14½	11½		
Urea	9	7	17	13½		
RESIN INGREDIENTS						
Hexamethylene	11	9	14½	11½		
Pentaerythritol	7	5½	14½	11½		
Phthalic anhydride	11	9	14½	11½		
MISCELLANEOUS						
Cornstarch	5	4	11	9		
Sulphur	–	–	11	9		
White dextrin	–	–	12	9½		
Coal dust, high volatile	10½	8½	15	12		
Petroleum pitch			11	9		

Notes

1. The 'Maximum recommended O_2 per cent' applies only to maintaining an inert atmosphere for protection against unexpected or unoikely sources of ignition. Much higher factors of safety are required where sources of ignition are deliberately applied. (continued on page 260)

Notes for Table 10.6

2. In the furnace test dust clouds of Zr ignited in CO_2. During heating for several minutes, undispersed layers of the following metals ignited (glowed) in CO_2: Mg, Zn, Mg/Al, Ti and Zr. Visible burning of dust layers was also observed in N_2 with powders of Mg, Mg/Al, Ti and Zr.

3. This table is an extract from NFPA Standard for Inerting for Fire and Explosion Prevention (No. 69). Copies of the complete text may be obtained from the National Fire Protection Association, 60 Batterymarch Street, Boston 10, Massachusetts.

apparatus. External explosions usually occur either when a pocket of hot dust, probably undergoing rapid oxidation, is scattered by a sudden rush of air, or by a local or internal explosion. An explosion stirs up dust deposits from ledges, the tops of plant and equipment or roof joists ahead of an advancing flame front. Coal dust explosions in mines often start as local, fairly mild fire-damp (methane) explosions, which stir up dust locally. These explosions, in turn, displace even more dust as they advance through workings and roadways and so gather tremendous force.

Burgoyne and Cohen[18] found for certain droplet clouds (aerosols) that very small drops (about 10 μ) volatalised ahead of an advancing flame front so that the lower inflammable limit was similar to that for vapour. Above 10 μ, they found that flame propagation depended on the ratio of droplet diameter to the distance between drops. The droplets vaporised and continued to burn behind the flame front. The lower flammable limit and flame temperature fall because the concentration near the drops is higher than elsewhere.

10.15 Insulating oil in transformers and switchgear etc[19]

When sparking or arcing occurs under oil, some of the oil which should be a blend of saturated hydrocarbons breaks down and inflammable vapour is produced. This may arise in several ways.

In the first place, depending on the current broken and the time it takes to extinguish the arc; a substantial gas bubble is always formed in a circuit breaker clearing a fault. If the breaker fails to clear the arc, the evolution of gas continues until the circuit is interrupted elsewhere. There is a similar, though usually smaller, evolution of gas at every operation of an oil immersed tap change switch or controller. These effects are normal and to be expected. Gas may also be evolved by tracking on oil covered insulators, which may also include some gas from the breakdown of insulating material.

An important, though fortunately not common, source is the breakdown of insulation on core bolts or between turns on the windings of

transformers, with continuous or intermittent sparking or small arcing. More common is poor contact in oil-immersed switches and circuit breakers caused either by surface irregularities formed by previous arcing at make and break or decrease of pressure by slackening of bolts or ageing of springs.

The composition of arc gas, which varies somewhat according to the source of the oil and other variables is approximately as follows

> Hydrogen 70 to 80 %
> Acetylene up to 12% (one investigation gave 22%)
> Methane 3 to 10 %
> Ethylene 2 %
> Other hydrocarbons and CO – small amounts.

Where there has been a core bolt failure in transformers it has been found that the flash point of the oil had fallen considerably compared with new oil. After a number of unexplained circuit breaker failures it was found that the flash point of the oil in other breakers had fallen from a nominal value of 145°C to 52°, 90°, 93°, 85°C. In one instance, in a circuit breaker, oil has been found, on warming, to give off dissolved gas containing 70 % hydrogen. Such gas which accumulates in an air cushion above the oil may burn in spite of the cooling effect of the oil on the rising gas bubble, but this does not always occur in a sealed enclosure with a limited amount of oxygen available.

After severe faults, transformer tanks frequently fail at the joints and a jet of oil is ejected which ignites on contact with air. On one occasion when a short circuit between turns had occurred, the top plate was slightly bulged and from a narrow gap between tank and lid a jet of flame emerged which burned paint off a door twenty feet away. Other failures on small transformers have shattered oil-filled insulators on adjacent large units of several hundred or thousand kVA capacity.

Following reports of external ignition of vapour from an overheated oil immersed resistor, Dr Alsop of the Safety in Mines Research Establishment heated a closed metal container of transformer oil with an immersion heater. After an interval of 90 minutes vapour escaping from a 0.006 in flange gap ignited spontaneously at a point some inches away from the tank, the temperature at the point of ignition being 293°C. This may well be associated with the cool flame phenomenon and perhaps the formation of peroxides.

The matters discussed above are relevant to the discussions of circuit breakers, flame proofing and its alternatives, and the hazards in electricity power stations in Chapters 7 and 12.

10.16 The spreading of inflammable liquids, gases and vapours

When an inflammable liquid escapes it is, in principle, possible to predict where it will go, but this is much more difficult for gases and vapours. *High flash-point liquids* may flow downhill, and consequently apertures in floors, underfloor ducts, and trenches must receive special attention. Apertures should be sealed and trenches or ducts should be filled with sand or alternatively interrupted by barriers. Where large volumes of liquids are involved essential floor openings may be protected by curbs, and stairheads by sills, so that each floor level becomes a large tank or several smaller ones. These units should be capable of retaining the contents of all vessels (and pipes if appropriate), with direct drainage to safe sumps, i.e. ones where a fire would not have serious consequences and with means of immediately quenching any flames if possible. Large tundishes for collecting drainage below storage and reaction vessels and small ones below valves, gauge glasses etc, leading to a sump, are also valuable.

The means of disposal of a liquid depends largely on its flashpoint. If this is high, as for transformer and switch oil, the sump may be immediately below the apparatus. If the sump is filled with clean pebbles or washed gravel of sufficient depth to hold all the oil with a foot or more to spare, this will usually prevent ignition or stifle flames. This method is standard practice in *outdoor* electricity stations and is often supplemented by automatic sprinkler or other protection (Chapters 7 and 12).

[*Note.* Attention should be given to the venting of drainage sumps and vessels which may contain volatile inflammable liquids. If a vapour/air explosion occurs the walls or roof may be shattered. Explosion relief, discussed in *'Dust Explosions in Factories'* (SH & W booklet No. 22, New Series) is relevant. In many circumstances a pebble filled tank open to the sky may be best, or with a *very light* roof which can be blown off harmlessly if the situation admits this. In enclosed situations provision for an emergency blanket of foam or an inert atmosphere would be an advantage. With a rapid rise of pressure even a large vent gives only limited pressure relief. The mass of the air in a large chimney has considerable weight and inertia and may not provide a good safety valve.]

When the flashpoint is low (below, say, 85°F (30°C)), the probable presence of vapour must be considered. It is often believed that special precautions are not required above 75°F (24°C) which is the limit under the 'Petroleum Storage Regulations' in Great Britain; but higher ambient temperatures are not uncommon. Examples of flashpoint are given in Table 10.1.

When the flashpoint is below the ambient temperature there will be a danger of local pockets of explosive atmosphere. There have been several investigations of the extent to which such concentrations will spread. These are largely based on the work of Sir Graham Sutton[20] on

atmospheric turbulence. The dispersal of a gas or vapour depends on the density. The following are some typical values:

Air 1.00

Methyl alcohol	1.11	Benzene	2.77
Methyl ether	1.59	Pentane	2.48
Diethyl ether	2.56	Petroleum ether	2.50
Ethyl alcohol	1.59	Petroleum spirit	3 to 4

As these vapours are heavier than air they will tend to flow to the ground. However, methyl alcohol vapour is of similar density to air and thus will mix fairly freely. In still or fairly still air, the others will tend to collect at ground level and spread out, mixing only slowly, and will flow along below-ground ducts and trenches and into pits and basements. However, if the vapour is above the air temperature the vapour will fall more slowly or even rise until it cools or mixes with the air.

If there are air currents, these will cause turbulance, which will increase with the velocity. This will cause rapid mixing through an expanding volume of air in a similar manner to that in which the smoke from a chimney forms a diffuse trail down wind. As the volume or trail expands, the concentration is reduced so that, if unconfined (as in the open air), there is a distance beyond which the concentration falls below the lower limit if flammability.

Liquid entering a closed vessel displaces its own volume of saturated vapour and on this basis Katan[21] using the results of Sir Graham Sutton, determined the minimum safe distance downwind from the filling aperture of an aircraft petrol tank while it is being filled with aviation spirit. The calculations appear to be in agreement with experimental values and the danger area is greatest when there is a strong temperature inversion but decreases with increasing wind speed. For petrol, Katan gives the equations:

For normal conditions

$$\text{minimum safe distance} = \left(\frac{52.5Q}{u}\right) 1/1.81 \text{ inches}$$

or, with strong temperature inversion   (10.2)

$$\text{minimum safe distance} = \left(\frac{215Q}{u}\right) 1/1.5 \text{ inches}$$

where Q = filling rate in gal/minute;
          u = wind velocity m.p.h.;
Steen covers somewhat the same ground in relation to measuring technique.

Table 10.7
EXTENT OF DANGER AREA AROUND FILLING POINT FOR NORMAL
METEOROLOGICAL CONDITIONS
Radius of danger area (ft) is given

Wind speed – u (m.p.h.)	Rate of filling – Q (gal/min)						
	20	40	60	80	100	120	150
0–1	3.7	5.7	7.1	8.3	9.4	10.5	11.8
3	2.1	3.1	3.7	4.6	5.1	5.7	6.5
6	1.4	2.1	2.7	3.1	3.5	3.7	4.4
10	1.1	1.6	2.0	2.3	2.7	2.9	3.3
15	0.9	1.3	1.6	1.8	2.1	2.3	2.7
20	0.8	1.1	1.3	1.6	1.8	2.0	2.3

Table 10.8
EXTENT OF DANGER AREA AROUND FILLING POINT FOR
CONDITIONS OF STRONG TEMPERATURE INVERSION
Radius of danger area (ft) is given

Wind speed – u (m.p.h.)	Rate of filling – Q (gal/min)						
	20	40	60	80	100	120	150
0–1	22	35	46	55	65	73	84
3	11	17	22	27	31	35	40
6	7	11	14	17	20	22	25
10	4.7	8	10	12	14	16	18
15	3.6	6	7	9	10	12	13
20	3.0	4.7	6	8	9	10	11

The following observations are relevant in considering the above tables.

1. In view of the fact that it is never convenient and often not possible to define wind direction accurately, the maximum distance down-wind within which danger exists should be regarded as extending in all directions, not only down-wind. Thus any distance quoted in the above tables for certain conditions should be regarded as the radius of the circular danger area for those conditions; the filling hole (or vent in the case of underwing filling) being the centre.

2. All distances quoted are likely to be substantially greater than those obtaining in fact for the following reasons:

(a) the air expelled from the tank may not be saturated;

(b) the vapour may be dispersed by chance draughts (particularly those caused by the movement of men and equipment), over and above the dispersion due to wind;

(c) some of the petrol vapour may escape through subsidiary channels, e.g. vents.

3. Whereas the values given in Table 10.8 should be employed if there is any doubt as to weather conditions or in any general approach to the problem, it must be remembered that the distances quoted are for the meteorological state of strong temperature inversion; such a condition is not very common, and almost invariably obtains at night. Since a factor of at least four times is involved, the

(continued on page 265)

Continuation of footnotes to Tables 10.7 and 10.8
usual danger radius for normal conditions will be given in Table 10.7, i.e. about a quarter that given in Table 10.8 for any given values of *u* and *Q*.

4. Strong temperature inversions are, moreover, associated with very low wind speeds; consequently the values in Table 10.8 for the higher wind speeds do not correspond to any practical conditions.
(From Fire Research Technical Paper No. 1). Reproduced with the permission of the Controller, HMSO)

Without mixing, however, heavy vapour may flow along the ground for considerable distances in a fairly compact stream. The following are examples from my own experience.

Petrol vapour from spillage at a petrol pump flowed across a garage floor into an inspection pit covered with heavy timber baulks. There was an explosion when an electric hand lamp was dropped and broken and the timber baulks were flung into the air (see section 10.10 above).

A reaction vessel was overfilled with ethyl ether which formed a pool on the floor. Vapour flowed about 20 ft to a doorway, out onto a roadway and along at right angles for 50 or 60 ft, and was ignited at a pump motor; the flame flashed back to the reaction vessel. There is a case on record where vapour from a vessel being heated apparently flowed out of the building to a cottage many yards away, and was ignited by an open fire, and flashed back.

In buildings the spread of liquid and heavy vapour may be prevented by sills, retaining walls or bunds wherever substantial quantities are handled or stored. In the open, if steps can be taken to increase air turbulence; this will usually reduce the area of risk.

Buildings containing potentially dangerous substances should be segregated and doors should not be opposite one another across a roadway or passage but staggered as far as possible. By taking such steps the area within which expensive safe electrical equipment may be necessary is substantially reduced. When all possible precautions have been taken to reduce the extent and area of risk there will remain situations where electrical precautions, such as the use of flameproof or intrinsically safe apparatus, are essential. The application of these is discussed in Chapter 11.

Evaporation of spilled petrol is discussed in relation to underground car parks in Post-War Building Study No. 28[22]. The authors calculate that the time required to fill the whole of the air space with the minimum explosive petrol vapour and air mixture would be

$$t = 0.017 \ h/n \ \text{or} \ h/n \ \text{minutes}$$

where t = time in hours;
  h = height in feet;
  n = fraction of floor area covered by liquid petrol.

On this basis the time required if h = 10 ft and n = 0.1, is 1.7 hours.

A more realistic assessment is that if the concentration decreases uniformly (monotonically) from the floor upwards the minimum time will be when the concentration is uniform vertically, and this will probably arise first immediately over the puddle, so that $n = 1$ effectively. Then, if the lowest piece of electrical equipment is at 4 ft from the ground:

$$t = 0.068 \text{ hours or about 4 minutes.}$$

Because of stray air currents the vapour may not actually reach the equipment, but according to this theory, the above is the minimum time required for danger to arise. If the height had been 1 ft representing a light switch above a work bench the time might be only one minute. I have investigated several such incidents.

These calculations give an indication of the order of magnitude of the problem, but they must be treated with caution until we have more experience in the field. Conditions are very variable and air currents unstable. It must be emphasised that they apply only to relatively small and restricted leaks and spillages. Major failures raise quite different considerations, as is indicated below (see section 10.17).

10.16.1 Evaporation

Evaporation from the surface of a volatile liquid will continue until the concentration immediately above the surface reaches a partial pressure equal to the saturated vapour pressure corresponding to the surface temperature. This is tabulated in a number of reference works; at boiling point it equals the atmospheric pressure.

Air movement, particularly turbulence, removes vapour from near the surface and prevents stability being reached. On the other hand the latent heat of evaporation is usually drawn almost entirely from the liquid and reduces its surfaces temperature and thus decreases the saturation value and thereby the rate of evaporation. In most liquids, this causes a negative temperature gradient, i.e. the coldest and therefore the heaviest liquid is at the surface (water near freezing point is an exception), this is unstable and sooner or later convection will allow it to sink. Anything which will restrict liquid convection, or turbulence, will reduce the rate of evaporation.

10.16.2 Liquified petroleum gases (including bulk methane and pentane) [25]

If we now turn to liquified gases whose boiling temperatures are below the environmental temperature, entirely different conditions prevail.

Among these, liquified petroleum gases are of great industrial importance, particularly methane, pentane, propane and butane (commercial products whether sold as propane or butane are to some degree mixed). Extreme examples of liquified gas are air, oxygen and hydrogen. These are only liquid at very low temperatures, but liquid oxygen, for example, is transported in large spherical insulated containers, without refrigerating equipment, by allowing free evaporation through a vent. The absorption of latent heat must be balanced by inward leakage of heat through the insulation and this keeps the loss by evaporation to a minimum without excessive internal pressure building up.

The boiling point of commercial butane is about $-11°C$ so that it can be stored without refrigeration at moderate pressures. The boiling point of commercial propane is about $-40°C$ so that it must either be stored refrigerated, or cooled by free evaporation (as for oxygen), but refrigeration is, I believe, only economical for large quantities (e.g. 2000 tons or more).

Propane and butane are heavy gases (see above) and the vapour will tend to collect at ground level and must be kept out of ducts, trenches, basements, etc in the event of leakage. The drainage of liquid should be away from storage tanks so that if there is ignition the tanks are likely to be clear of the fire. On an open site away from likely sources of ignition it is possibly best to have low bund walls so that vapour will be dissipated by turbulent mixing with the air, a windy side being best.

Quite different considerations may arise however from a major leakage of gas or a highly volatile liquid such as from a split tank or the failure of a large pipe. Also precautions must be taken to prevent manageable fires from becoming calamities — see Chapter 2 and below.

10.17 The Flixborough explosion

In 1974 a very serious vapour explosion occurred in a chemical plant at Flixborough following a major pipe failure. Fortunately the works was in a sparsely populated area near the banks of the River Trent. In spite of this however, extensive damage was done to several villages and in the town of Scunthorpe several miles away. By any standard the installation is seen with hindsight to have contained a 'calamity risk' as defined in Chapter 11. Electricity played no part in this incident, but in the absence of other sources of ignition an electric spark could initiate such an explosion when a fracture releases an enormous volume of explosive gas or vapour.

The government ordered a public enquiry into the lessons which may be drawn from such an event (a summary of the Report is given in the Appendix on page 356). I have discussed the spread of small

amounts of flammable liquids and gases but at the present time adequate theories and data do not exist on how to handle such major escapes apart from standard precautions to prevent their occurrence, and making provision for all electrical supplies to be cut off immediately over a wide area.

Apart from this the 'Advisory Committee on Major Hazards' was set up by the Health and Safety Commission which published its first report in 1976 (HMSO) which outlines a policy of administration and planning. No doubt later reports will deal with the technical aspects of the problem.

10.18 Permanent gases

The behaviour of permanent gases depends on their density. Typical examples are given below:

CLASS A		CLASS B	
Air	1.00	Ammonia	0.59
Oxygen	1.10	Methane	0.55
Nitrogen	0.97	Hydrogen	0.07

CLASS C		CLASS D	
Acetylene	0.91	Butane	2.05
Carbon monoxide	0.97	Propane	1.56
Ethane	1.04		

Class B gases will rise rapidly and collect at ceiling level and it is important that the ceiling should be ventilated at its highest point and not act as an inverted tank in which gas can collect. In the open air these gases will usually rapidly disperse without danger except in the immediate neighbourhood of the leaks, unless the escape is large. Only safe electrical equipment should be used at ceiling level. On one occasion an escape of town gas in a regulator room collected overnight and was ignited when an operator switched on the light in the morning. He was blown across a yard but not seriously injured.

Class C gases will usually mix freely with air by eddy diffusion without much tendency to rise or sink. For this reason they will not tend to collect if leaks are moderate and there is good ventilation. It is easy to calculate the number of air changes an hour necessary for adequate dilution but allowance must be made for lack of uniformity.

Class D gases will tend to sink to the floor in still air and the conditions will tend to resemble those for vapours discussed above.

Oxygen presents a special hazard. Although it does not burn in air it should be treated as a highly combustible gas because many substances such as used in electrical insulation which are non-flammable or only burn slowly in air will burn violently in oxygen. Because oxygen is slightly heavier than air, slow leaks near ground level in calm conditions may tend to collect in depressions and having no colour or smell may be a very insidious hazard. On one occasion an oxygen lance or welding torch was left in the 'dish' formed by the end for a large cylindrical vessel and it filled with oxygen to the brim. When later the operator attempted to relight the lance or torch his clothing burned fiercely and he died. The ignition might have come from an electric portable tool if oxygen had been drawn over heated insulation by the ventilating fan.

10.19 Explosives

Explosives may be divided into four classes: propellants, blasting explosives, military high explosives and detonators or, more correctly, initiators. A detonation occurs when a pressure or shock wave is propagated through the explosive at very high velocity, so that the whole mass explodes virtually simultaneously. The velocity ranges from about 3000 to 9000 m/s, and the temperature it generates may be as high as $10\,000°C$.

Broadly speaking, propellants will not detonate in normal use; they must be ignited by a flame. High explosives for military purposes will only detonate as a result of a very violent shock; a shell has to be accelerated along a gun barrel and possibly penetrate a foot or more of armour plate without detonating. They are therefore comparatively safe to handle. TNT burns relatively slowly with a smoky flame when it is not confined, the total release of energy being less than for an equal weight of fuel oil.

Some blasting explosives detonate more easily and are more dangerous to handle. A stick of blasting gelatine, for example, if ignited by a flame, will start to 'burn', but this may suddenly turn to detonation. The special blasting cartridges used in coal mining detonate, but they have been tamed by adding material, such as common salt or sodium bicarbonate, either to the mixture or as a sheath, which cools or deactivates the gases produced, so that they will not ignite a methane–air mixture. Between these extremes, there is a whole range of degrees of sensitivity and danger in handling.

Initiating compounds and substances are highly unstable and detonate easily. Some are very sensitive, so that, whereas the energy required to ignite, say, TNT, is of the same order as most hydrocarbons, i.e. a few hundredths or tenths of a joule; copper acetylide requires only

about 20 erg, and basic lead trinitroresorcinate 30 erg. Most initiators are less sensitive than copper acetylide, but considerably more so than TNT. Initiators are mostly weak explosives; i.e. the total energy output is relatively low, but the rate of release is very high. To detonate TNT, an intermediate detonating substance, more sensitive than TNT, but more energetic than the initiator, may be required; this is called a gaine. To fire propellants, a hot flame is required, which may be produced by a small intermediate charge of gunpowder.

Any fire in an explosives works is obviously dangerous. The principal hazard in many factories is likely to be from dust, so that the use of dust-tight electrical equipment is usually essential. Now that a testing specification is available, this can be ensured, but an independent testing and certifying agency is highly desirable; and the construction must be robust.

TNT (trinitrotoluene), picric acid (trinitrophenol) and similar substances are made under conditions similar to those in the fine-chemical industry, and, since they are made from flammable raw materials, particularly benzene, a flameproof installation is necessary where a flammable atmosphere may arise. Picric acid and some other substances, however, form highly unstable salts with the heavier metals, which must be excluded from all paints and enamels in places where they are made or used.

Dusty processes, such as filling sporting cartridges and making fireworks, are usually carried out in small buildings which may be protected by screens or mounds. Electrical equipment must be dust-tight, or better still excluded, lighting being through windows or roof lights.

Special precautions are necessary with more sensitive detonating compounds, such as some blasting explosives, and extreme care is necessary with initiating compounds. With some of the latter, the static electricity generated by sliding powder down a chute may cause an explosion, and special precautions against static electricity are necessary, but these fall outside the scope of this review.

Blasting explosives are used, to some extent, in general industry e.g. for demolition, breaking up heavy steel scrap and demolishing slag heaps. The primary electrical hazard arises from the electrical exploders used, but prevention is largely a matter of common sense and following, so far as is applicable, the practice developed in mines and quarries.

Common sense must be used in storing and handling cartridges and packed explosives, and, if a store is illuminated, a high standard of electrical installation is necessary, preferably dust-tight.

Explosive-manufacture and filling factories, including those for civilian purposes, such as fireworks, railway fog signals, marine rockets and blasting explosives and detonators, are subject to the Factory Electricity Regulations in Great Britain. Private establishments other

than for military purposes, are also covered by the Explosives Acts and are inspected by HM Inspectors of Explosives. Their safety depends primarily on this, but a safe electrical installation is one of the necessary precautions. Nobody should undertake electrical-installation work in an explosives works without specialist advice, and the above notes are intended only to indicate the type of hazard which may be met, they are not a complete technical guide.

10.20 Radiation fields[26]

In a strong radiation field, such as a radar beam, a conductor may pick up sufficient energy to cause the ignition of gases or sensitive explosives, particularly if it happens to resonate at the radio frequency. Radiation or static induction from thunder clouds and lightning discharges have caused a number of explosions. It may also be possible for induction from power lines to do so.

These matters have been covered by BS 4992.

10.21 Electrostatic hazards[28]

In the interior of any body, the electrostatic and quantum forces are, on the average, in balance. At the surface, therefore, there must be a sharp transition, and, in particular, short-range forces and potential gradients appear. As a consequence, when two bodies with different physical properties come into close contact, one body tends to acquire electrons or ions and the other to lose them. When the bodies are separated, if they are insulators, their surfaces retain bound charges. Cleavage of a crystal, a difference in temperature or strain may be sufficient to cause this interchange. The separation causes a difference in potential, which may become very large as the separation increases; but, because the surface charges are very small, the available energy is also fairly small.

At the surface of a polar or dissociable liquid, or an insulating liquid containing dissociable or polarised impurities or dissolved molecules, these unbalanced forces cause an electrical double (Helmholtz) layer at any surface. This leads to separation of charge whenever there is splashing or bubbling, and, in particular, a flow of liquid produces a slip in the double layer (Figure 10.2), which produces a relatively large separation of charge. Differential freezing of ions also causes separation at low temperatures.

As a result of these and some other effects (such as piezo-electricity), charges are always formed whenever solids, liquids, dusts or droplets

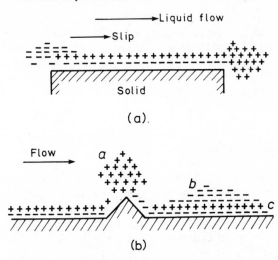

Figure 10.2 Electric-charge separation
    (i) Charge separated by slip of double layer
    (ii) Transient charges liberated at discontinuity
        (remixed by turbulence)
    (a) Free positive charge swept away
    (b) Free negative charge released by formation of
        new positive charge
    (c) New free positive layer
                                    (Brit J Appl Phys 1953)

are moved or handled. The only reason we do not notice them is that a very small degree of conductivity is sufficient to dissipate them. It can be proved that bulk charges will disappear to the surface if the conductivity is greater than about $10^{-10}\,\Omega/cm$. They will then be transferred to the containers, and, if these are metal, semiconductors or poor insulators, thence to earth. If, however, the containers are insulated from earth, there will be trouble. The history of an unwanted bulk charge is shown in Figure 10.3. If the resistance to earth is less than about 10^8 or $10^9\,\Omega$, the charges will disappear so quickly that they are unnoticed.

The energy produced by even a small flow of liquid is sufficient to cause ignition of sensitive explosives or inflammable vapours, such as petrol, and this may arise from filling a bowl or squeezing out a washing leather or rag. Handling powders in bulk also produces dangerous charges. All that is necessary to dissipate them safely is to reduce the bulk resistivity well below $10^9\,\Omega/cm$, and earth and bond together all the containers. The flimsiest earth connection is sufficient, in theory, but it must be made substantial to stand up to the mechanical wear and tear of industrial life.

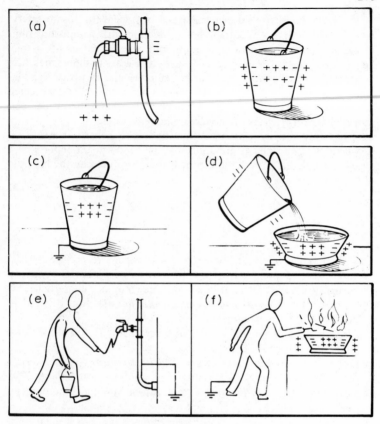

Figure 10.3 The history of a bulk electric charge

    (a) Flammable liquid becomes charged by flowing through tap; an equal and opposite charge is left behind

    (b) The charged liquid induces equal and opposite charges on the walls of the bucket in which it is collected

    (c) The bucket is earthed by placing it on the ground and loses its outer charge; although both the bucket and the liquid are charged, they are now neutral as a whole

    (d) The liquid and its charge are poured into a basin, and the process of inducing charge is repeated; the exterior charge on the bucket is left behind

    (e) An operator in rubber-soled shoes shares the charge with the bucket and a spark passes from his hand to an earthed fuel tap causing a fire

    (f) Another person without rubber-soled shoes (who is 'earthed') puts his hand near the surface of the charged liquid and a spark passes causing another fire; this has in fact caused many accidents and earthing the basin makes very little difference

(Reproduced by permission of the Controller of Her Majesty's Stationery Office)

In most situations, surface leakage is sufficient to ensure safety, but, in dry or frosty weather, the conducting surface-moisture films break up. Thus a reduction in air humidity from 60 to 40 % may increase the surface resistance of a sheet of glass a million times. The greatest hazards therefore occur in hot summer and in frosty weather. Out of doors, the moisture film breaks up into ice crystals, and, indoors, cold air from the outside, which may initially be of high-percentage humidity, is very effectively dried by warming.

It has only been possible to outline the problem here. It is, in fact, very complex. Further information may be found in the reports on a series of conferences on Static Electrification organised by the Institute of Physics.

Acknowledgement

This chapter derives mainly from a review paper *Electrical Safety in Industry* which I prepared for the Institution of Electrical Engineers (Proc IEE Vol 117, 1970) but I have also drawn upon other articles, i.e.

Insulating oil in relation to circuit breaker failures, Jour IEE Vol 90 Pt 2 (1943)

Electrical equipment in relation to areas of low fire risk, Electrical Supervisor (Dec 1967)

Electrical hazards in danger areas. Classification of risks and certification of equipment. IEE Conference on Electrical Safety, 1971

Automatic fire detection systems. Industrial Systems & Equipment, 1967

References

1. Lawson, D.I. and Fry, J.F„ 'Fires of electrical origin', *Proc IEE* 104 A (1957)
2. Gosland, L., 'Age and incidence of fires in electrical installations', *Proc IEE* 103 A (1956)
3. *Fire Protection Handbook*, National Fire Protection Association, Boston, USA
4. Morgan, J.D., *Principles of ignition*, Pitman (1942)
5. Jones, E. Taylor, *Induction coils and their applications*, Pitman (1932)
6. Bradley, J.N., *Flame and combustion phenomena*, Science paperbacks, Chapman & Hall
7. Coward, H.F. and Jones, G.W., 'Limits of inflammability of gases and vapours', *Bulletin 279*, US Bureau of Mines
8. Sloane, R.W., 'Ignition of gaseous mixtures by the Corona Discharge, *Phil. Mag.* (19 May 1935)
9. Sutcliffe, R.A., *Private communication*

10. Klinkenberg, A. and Van der Minne, J.L., *Electrostatics in the petroleum industry*, Elsevier (1958)
11. Titman, H., 'A review of experiments in the ignition of inflammable gases by frictional sparking', *Trans. Inst. Mining Engineers* **115** (1955/56)
12. Roberts, A.F. and Clough, G., 'Thermal decomposition of wood in an inert atmosphere', *Int. Symposium on Combustion 1963*, Academic Press, New York
13. *Studies of growth of fire*, Fire Protection Association
14. Brown, K.C. and James, G.J., 'Dust explosions in factories – a review of the literature', *Safety in Mines Research Establishment Report 201* (1962)
15. Rarbash, D.J., *Fires involving dusts*, Fire Protection Association Technical Booklet No 5
16. *Dust explosions in factories*, Safety Health & Welfare Pamphlet, New Series, No 22, HMSO
17. Palmer, K.N., *Dust explosions and fires*, Chapman & Hall (1973)
18. Burgoyne, J.H. and Cohen, L., 'The effect of drop size on flame propagation in liquid aerosols', *Proc Royal Society*, **Series A No 462 225** (1954)
19. Cooper, W. Fordham, 'Insulating oil in relation to circuit breaker failure', *Jour IEE* **90** Pt 2 (Feb 1943)
20. Sutton, O.G., *Atmospheric turbulence (Scr)*, Methuen's Physical Monograph
21. Katan, L.L., *The fire hazard of fuelling aircraft in the open*, Fire Research Association Technical Paper No 1, HMSO
22. *Precautions against fire and explosion in underground car parks*, Post-War Building Studies No 28, HMSO
23. *Industrial solvents and flammable liquids*, Fire Protection Association Pamphlet No 24 (1954)
24. Durans, A., *Solvents*, Chapman & Hall, London (1946)
25. *The bulk storage of petroleum gas in factories*, Safety Health & Welfare Pamphlet, New Series No 30, HMSO
26. BS 4992: 1974: *Protection against ignition and detonation by radio-frequency fields*, British Standards Institution
27. *Data published by the Fire Protection Association*
28. 'Report of Conference on Static Electrification', *Suppl. No. 2, Journ. of Applied Physics* (1953). Subsequent conferences have been held relating to this subject and the proceedings have been published by the Institute of Physics.

Chapter 11

Electrical Equipment for Danger Areas

11.1 Introduction

In the greater part of this book I have dealt with general principles and mentioned rules and regulations only incidentally, but the practice of electrical installation work in danger areas is so closely related to national standards that it can only be understood against that background. At the present time these standards are subject to considerable discussion and reappraisal, both from technical aspects and also because attempts are being made to harmonise the practice of the leading industrial countries.

In the past there have been two main schools of thought, the British and the German. It seems that USA practice is on the whole similar to the British but is more detailed and restrictive on some points, but USA policy is also less clearly defined because different states have different rules. Russia and East European countries, in the main, approximate to German standards, but will also accept British standards. The Commonwealth (except Canada) mostly follows British practice. Western Europe is divided between British and German with some states accepting both, but the underlying scientific principles are fairly well established. Because of the current confusion which stems from the attempt to reconcile specifications based on very different national design philosophies, it is necessary to explain in considerable detail how the present position has arisen. A common EEC policy is being evolved.

The information in this chapter is not intended to lay down a standard for the design of safe equipment, or for its detailed application. Reference should be made to the very extensive literature on the subject and also to the legislation and current standards applicable to a particular installation. The discussion is intended to help the non-specialist to understand the scientific background and thus to make

better and more economical use of the equipment available, but not as a substitute for expert professional advice.

11.2 Flameproof electrical equipment, segregation and pressurisation

11.2.1 General principles of flameproofing

Power circuits and apparatus which are necessarily exposed to flammable atmospheres may be placed in flameproof enclosures so designed that if an internal gas mixture is ignited for any reason, escaping gases are quenched, so that any external flammable atmosphere will not be ignited. This is normally achieved by providing all parts of the enclosure with wide, accurately machined flanges which are bolted together. When motion must be transmitted between the inside and outside (e.g. by motor spindles or operating levers), flameproof bushings or glands, designed on the same princile, are used.

In British practice there have been four main groups of flameproof equipment. Group I was confined to coal mining (which is not the concern of this chapter). Group II covered equipment for most industrial gases and solvents, whilst Group III was for a small number of gases which presented special difficulties, notably coal gas and coke oven gas. Group IV covered those gases and vapours for which it was not possible to make flameproof equipment commercially, e.g. acetylene, carbon disulphide, ethylene nitrite, hydrogen and blue water gas. It is possible, however, that in the future low volume flameproof enclosure may become available for hydrogen. These groups have now been replaced by I, IIA, IIB and IIC, but the original terminology is retained here because this was used for research work on this subject.

11.2.2 The experimental basis of flameproof designs

In the early days of flameproof electrical equipment for coal mines, flange gap clearances were considered as safety devices to reduce internal pressure, and considerable ingenuity was used in prototype designs to ensure that exactly the right gap was maintained. How far these and other means of venting were ever used is doubtful, they certainly did not survive long in practice and 'rough' machined metal to metal joints, bolted tight and without deliberate pressure relief became the rule in industry generally, the permitted gap being for the most part

a manufacturing tolerance. Rough machining here implies a coarse ground surface, not an inaccurate one.

In the early 1930's, when the need for flameproof equipment for use in industrial situations became urgent it was found that the permissible gap clearance provided a valuable method of grouping gases and vapours for test, certification and manufacture. From this has evolved the new grouping that is now in use.

Sealing the flanges with an external rubber band or adhesive tape has been found to be particularly dangerous, but it has for many years been a common practice to cover the matching surfaces with a film of grease before bolting up. This provides an effective seal which will keep moisture out and flame in. It has been suggested that with high internal pressure, particles of burning grease could be ejected, but many industrial greases are heavy-metal soaps which will not burn. In any case, with such small clearances, it is easy to calculate that the viscosity of even thick oil, let alone grease, is sufficient to prevent it moving perceptibly during the short time of active burning.

Paint has also been suggested as a sealant but this can harden and become detatched in patches thus upsetting the effective gaps; it is not, therefore, recommended. Attention has also been given to the possibility of using gaskets. The general conclusion appears to be that while they may be effective in principle, possible deterioration in use or during replacement, or lack of replacement, is a serious drawback and they should not be used unless and until they are specifically covered by specification and test certificate.

It seems that it was originally believed that the effect of the flanges was to cool the escaping gas below ignition temperature; more recent researches, however, suggest a more complex picture. It has long been known that flame, or at least incandescent gas, may escape through a 'safe gap' without igniting an external flammable atmosphere. In a paper in 1953, H. Smith sutdied the cooling of gases on passing through the gap but 'did not consider that the transmission of flame was influenced to any great extent by this effect'.

H. Phillip [5a], [6a], in two papers, concluded that the flame, i.e. active burning, is extinguished (deactivated) in the gap but hot combustion products are ejected from the gap and mix with the surrounding atmosphere, giving a jet of hot flamable mixture. External ignition may then occur unless the temperature is initially too low, or more commonly, turbulent mixing cools the jet externally before sustained combustion is established.

More recently Phillips has given a quantitative theory and has found good agreement between calculated and experimental safe gap values. This theory, that the probability of external ignition depends to a material extent on the velocity, turbulence and rate of cooling of the

escaping jet, provides a qualitative explanation of the observation that safety depends critically on a free escape. That is to say, equipment which would be safe in the open may be made unsafe if the jet impinges on a nearby barrier or by secondary (series) flanges, or obstructions within the flanges. These results place important limitations on the design and utilisation of flameproof equipment. That failures have not occurred is probably because in practice flanges are bolted tight. This is also a protection against the possiblity of hot particles (sparks) being ejected, particularly if conductors or other details are of aluminium. No doubt the same would apply to light alloys e.g. of magnesium and cadmium.

11.2.3 Relevance of safe gaps

If, in fact, flanges are almost universally bolted up tight, one may ask whether the grouping by flange gaps continues to be a controlling feature in design, except as a criterion of severity for testing, and in fact the great majority of new industrial equipment, excepting motors, was certified for Group III.

For testing, the grouping has been useful as a criterion for selecting suitable test atmospheres and it has been found that, even with the heavy construction necessary for flameproof equipment, in the event of an internal explosion the flanges may flex slightly between the clamping bolts and thus increase the effective gap (if any).

Rotating equipment needs special consideration. Statham and Wheeler[1] stated in 1930 that the turbulence of an inflammable mixture, such as produced within a motor casing when the motor is running, has no appreciable effect on the maximum pressure developed but has a considerable effect on its rate of development. American work on large machines has suggested that this matter may be much more important than was once believed, and full-scale research has been undertaken. Flameproof sealing of the shaft clearance of motors, which cannot be bolted tight in the same way as flanges has also presented difficulties. Therefore, for some gases and large motors it is necessary to use a labyrinth gland to prevent the passage of flame. In prototype and routine testing of apparatus, static as well as motors, testing on load is under consideration but it is not yet possible to say what the final outcome will be.

However, despite doubts and difficulties, in practice testing and certification based on a classification by permissible flange gaps has been remarkable successful and over a period of about 40 to 45 years of increasing use in industry, there have been very few failures. I can personally remember only two, and these were a result of structural weaknesses in very early designs and had nothing to do with 'gap theory'.

11.3 Intrinsically safe low-current circuits

11.3.1 General principles

In the discussion of electric spark ignition in Chapter 10 we saw that a minimum amount of energy must be supplied to a spark for ignition to occur. In instrumentation, telecommunication and control circuits, the danger arises from break circuit sparks, as in a switch or at a broken lead or an intermittent short circuit. If the supply is over approx 20 V there is a tendency to momentary arcing (which is very incendive) unless the current is limited to a very low value by non-inductive resistance in the supply circuit or otherwise. Apart from this, higher voltages are only permissible if the maximum current is very small, and currents may exceed a fraction of an ampere only if the voltage is very low. The stored energy of the broken circuit $(I^2 L + V^2 C)/2$ contributes substantially to the spark energy and must be kept to a minimum.

There has been a great deal of research on this matter in the UK carried out jointly by the Safety in Mines Research Establishment and the Electrical Research Association. This was conveniently summarised by Cartwright.[7] As with flameproof equipment, the possibility of danger is statistical and while calculated figures provide an indication of whether a particular circuit is likely to be safe, the final practical criterion of safety is a type test with standard break circuit apparatus. A detailed account has been given by Redding[8]. For the purpose of certification industrial gases and vapours were divided into 5 classes in BS 1259.

11.3.2 Particular problems — barrier units

Intrinsic safety was first developed by Professor Thornton following the investigation of a disastrous colliery explosion, which was believed to have been caused by a spark between the leads to a signalling bell. As a result, it was realised that some control must be exercised over bells, wiring and the sources of supply. When the principles were extended to telephone circuits it was found that, although individual manufacturers produced safe systems, if for example a bell was used with a magneto ringer made by a different manufacturer the result was not always safe. This led to the manufacturers producing agreed standards for various items so that, where necessary, these units could be combined in safety.

In the early 1950's a number of explosions occurred with industrial instrumentation and telecommunication systems and it was obvious that this problem would increase. A number of systems were tested and certified but with increasing instrumentation and automation it became

clear that to certify complete systems, which might mean a new certificate for each installation, was unworkable and attempts were made to produce standard sources of supply which could be used with a variety of types and combinations of apparatus.

Some progress had already been made in the production of 'safe coupling units' for telephones where only part of a telephone system need be in a danger area, although these were rather clumsy; but with

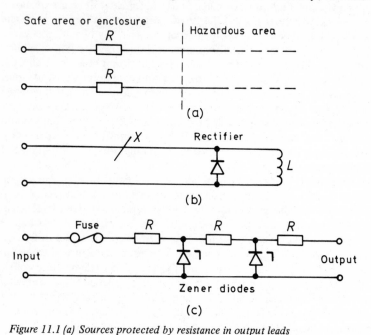

Figure 11.1 (a) Sources protected by resistance in output leads
(b) Inductive load protected by short circuiting rectifier
(c) Typical Zener diode coupling unit
Note. In (c) the fuse and resistances must be so proportional that in the event of an external short circuit the diodes are not destroyed by overheating before the fuse blows

the growing use of sensitive equipment and the development of industrial electronics it became possible to use circuits with such small currents that danger was virtually impossible provided the source of supply was 'safe' i.e. the maximum possible current input on short circuit was effectively limited and the interconnecting cable routes were non-conductive. A number of ways of producing safe systems were already known. Figure 11.1(a) shows a supply unit, to be placed in a safe area or flameproof enclosure where the current is limited by a resistor in each supply lead. Figure 11.1(b) shows a circuit with the

inductive impedance of an instrument or bell shorted by a rectifier; if the lead breaks at X the inductive energy is discharged through the rectifier instead of the break. The most recent development is the use of a Zener diode in place of the rectifier; Figure 11.1 shows this use. Applications of this principle are given in Redding's book[8].

There are, however, precautions which must be taken. The total stored (inductive) energy of the circuit must still be limited, which implies some limitation of cable runs, and great care must be taken to see that there is no possibility of contact or leakage between the safe and unprotected sidesof the barrier units.

The final vetting of such equipment should not be left to laboratory staff. For example, the whole safety of a unit might be lost if the casing would permit conducting (or even semi-conducting) dust to enter, or moisture or contamination by a corrosive atmosphere. If the whole unit is cast or encased in transparent water-repellent plastic this will be an advantage. In the case of thermoplastics the equipment should not be sited too near a furnace. The whole unit must be sufficiently robust and resistant to corrosion or contamination to withstand considerable abuse.

Statements in several papers and articles have implied that explosions caused by instrumentation have not in fact occurred, e.g. Elterton[10] in the symposium in 1962 wrote: 'H.M. Factory Inspectorate has no record of any explosion touched off by an instrument during post-war years'. Within my own experience the following incidents have occurred. A coke oven booster house was 'blown up' when leaking gas was almost certianly ignited by telemetering equipment. Another booster house was blown up by a spark from automatic telephone contacts. A small instrument house had the windows and doors blown out when a percentage oxygen recorder exploded and a similar explosion occurred in the casing of a ring balance flow meter.

There is always a hazard when gas sampling or pressure pipes must necessarily be brought into the casing of an instrument with electrical connections, including electrical clocks or motors for driving charts. In a segregated control room (see 11.5) two control panels were blown up because of gas leaking from the small pipes to pressure relays.

11.3.3 Revised grouping (BS 5345)

It had long been appreciated that there was a correlation between flameproof grouping under BS 229 and the classification for intrinsic safety under BS 1259, this was elucidated by Slack and Woodhead[9] from whose paper Figure 11.2 is taken. This correlation leads to the

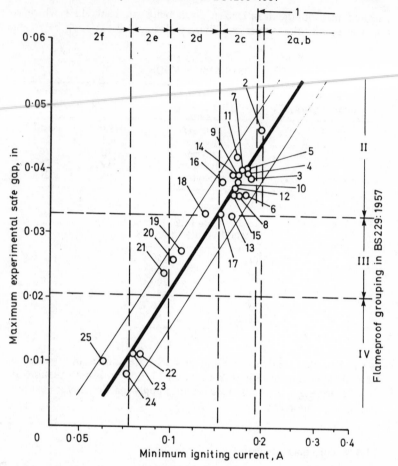

Figure 11.2 Correlation of maximum experimental safe gap with minimum igniting current, excluding the data for ammonia. Limits of 95% confidence shown (Slack and Woodhead[9])

2 Methane	14 Benzene
3 Chloroethylene	15 Carbon monoxide
4 Methyl acetate	16 Propane
5 N-butyl acetate	17 Diethyl ether
6 N-heptane	18 Buta-1 : 3-diene
7 Acetone	19 Ethylene
8 Methyl alcohol	20 Ethylene oxide
9 Pentane	21 Town gas
10 N-hexane	22 Water gas
11 Butane	23 Hydrogen
12 Cyclohexane	24 Carbon disulphide
13 Methylethyl ketone	25 Acetylene

revised method of classification given below in Table 11.1 which is associated with the temperature limits in Table 11.2.

Table 11.1
BS CLASSIFICATION

BS 4683 apparatus group	Representative gas	Former definitions BS 229	BS 1259
I	Methane	group I	class I
IIA	Propane	II	2c
IIB IIB	Ethylene	III	2d
IIC	Hydrogen	IV	2e
(not yet allocated)	Acetylene	IV	2f

Table 11.2
TEMPERATURE CLASSES

T1	450°C
T2	300°C
T3	200°C
T4	135°C
T5	100°C
T6	85°C

It will be noted that these are not precisely the same as in the original German temperature ratings (see section 11.9.6). In some situations it may be necessary to check which figures apply, particularly in relation to existing equipment.

11.4 Modern developments

An important development is the introduction of optical devices. Glass fibre bundles are already used to transmit information or pictures from otherwise inaccessible points. Recently, however, devices have been developed in which at some point 'information' is transmitted optically through a block or film of transparent insulating material and picked up photoelectrically on the other side. This seems an ideal barrier unit if the thickness of insulation is sufficient, since there is little possibility of electrical contact between the two sides, but of course such a unit must be protected from other light sources or spurious signals will be picked up and it will still be necessary to limit the energy in the 'safe' circuit.

Many solid state electronic components such as diodes are liable to damage by slight over-voltages; this is discussed in Chapter 9 'Switches

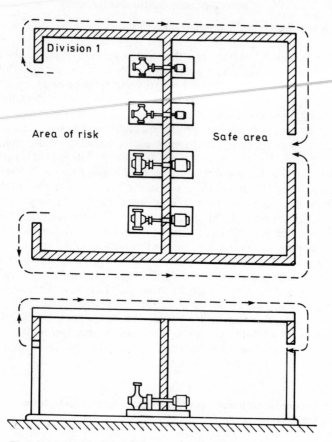

Figure 11.3 Segregation: application of physical barriers to prevent flammable concentration invading nonflameproof or nonintrinsically safe electrical apparatus.

The roof and walls are required to be made of noncombustible low-porosity materials. Within the area of risk, any electric apparatus should be flameproof, intrinsically safe or such other type affording an equal degree of protection. The shortest possible paths for flammable concentrations to pass from the area of risk to the site of nonflameproof or nonintrinsically safe electrical apparatus are shown in broken lines (CP 1003)

and control systems'. Some components are easily destroyed by over-current, and the voltage and current 'withstand' values must be respected. Connections spreading over a large area, e.g. several acres, are particularly vulnerable to quite small earth potential gradients which could arise from power faults, testing, remote thunderstorms, or even electrolytic effects in the ground.

11.5 Segregation and pressurisation

When flameproof equipment either cannot be used, is not available, or is excessively expensive, a common procedure is to segregate motors and control gear from the danger area and to drive plant by a shaft through a gland in the separating wall (Figure 11.3). Alternatively a positive internal vapour-free air or inert gas pressure is maintained in the casing, effectively preventing an internal explosive atmosphere.

It is also necessary to prevent back-circulation of air from the danger area (Figure 11.3), and to establish the internal pressure for a sufficient time to ensure that casings (and conduit if used) are scoured free of any explosive mixture before the power is switched on. If the internal pressure fails the power must be immediately cut off; automatic flameproof and intrinsically-safe relays are necessary for this purpose. The general requirements for alternatives to flameproof construction are discussed in BS Code of Practice CP 1003 Part 2.

Segregation and/or pressurisation is also used where it is impossible to make instruments or control apparatus intrinsically safe throughout. Thus the sensing element and its wiring may be safe, and connected by an intrinsically safe circuit, to a recorder, relay or controller incorporating non-safe elements, either outside the danger area, or in a flameproof enclosure within the danger area. Such an enclosure must be interlocked with an isolator in a separate flameproof enclosure to ensure that all pressure is cut off before the case is opened.

11.6 Oil immersed flameproof transformers; switch and control equipment and fusegear (see section 10.15)

Oil immersed equipment poses special problems. Flameproof enclosure has been defined as

'enclosure of electrical apparatus . . . that will withstand, without injury, any explosion of the prescribed flammable gas that may occur within it under practical conditions of operation *within the rating of the apparatus (and recognised overloads, if any, associated therewith)*, and will prevent the transmission of flame such as will ignite a prescribed flammable gas which may be present in the surrounding atmosphere' (BS 229 : 1957).

In the case of oil immersed switch and control gear the internal flammable gas may be 70 to 80 % hydrogen and one is left in some doubt how far it is practicable to make the equipment fully and effectively flameproof, particularly bearing in mind the conditions which arise when clearing a heavy short circuit, when even the best circuit breakers are often strained. After clearing a heavy fault, or several smaller ones, deposits of carbon on insulators have frequently — after time to settle

— led to insulation failure and switchgear explosions. It has also been demonstrated that internal faults in transformers and circuit breakers may lead to the ejection of vapour/air mixtures which ignite spontaneously on contact with the surrounding air at comparatively low temperatures. This places severe limitations on the siting of oil immersed flameproof equipment. For these reasons I would myself only accept oil filled apparatus in 'Remotely Dangerous' (Division 2) areas as described below, and only then if it was unavoidable. Fortunately dry type circuit breakers filled with sulphur hexafluoride are, or soon will be, available. This is a non-inflammable gas, with very good electrical properties, i.e. high dielectric strength and arc quenching ability. Other vapours and liquids have also been used or tried.

Dry type transformers have been known for many years. In a hazardous (flammable) situation they must be flameproof and because of the possible evolution of flammable gas from insulation under fault conditions, the use of inorganic (Class C) insulation is desirable. Somewhat similar but less severe problems arise with fusegear and air-break control gear. It is essential to protect such apparatus with back up h.r.c. fusegear which will nip off any short circuit in the first half cycle before a current value is reached which can cause any dangerous stress.

Back up protection of switch and control gear should, if at all possible, be located in a 'safe area' or protected enclosure. There is an important distinction between industrial and mining practice which should be noted. In coal mines circuits are (or should be) made dead before any work is done on electrical equipment. For this reason switch and fusegear has not had local back up isolation. This is not practicable in general industry and for this reason such equipment is provided with local isolating switches, which can be locked open, to facilitate maintenance. The isolating switch must of course be in a separate flameproof enclosure with no possible flame path between it and the isolated equipment.

Switch and control panels may fail through 'pressure piling'. This occurs when two enclosures are connected by a restricted passage; if ignition occurs in one of them a pressure wave precedes the flame and precompresses the mixture in the other, (similar to the compression stroke in a petrol engine) so that the pressure rise int he second is much more violent than it would otherwise have been. This can happen to some extent in a controller where the equipment tends to divide the space into two (or more) partially separated volumes, especially if they are unequal.

11.7 Summary

Where there may be an explosive atmosphere, local or general, there

have been a number of alternative means of obtaining safety. These
can be summarised as follows:

1. *Segregation*. This means the removal of all or part of the electrical equipment from the danger area and placing it in a safe
   situation.
2. *Pressurisation* (see section 11.5) has a limited application. It can
   be applied to controllers and instruments but entails considerable
   complications.
3. *Flameproof equipment* may be used for power circuits, but
   should not be used where a flammable atmosphere is always or
   commonly present under normal working conditions. There are
   considerable limitations to its application to oil-immersed equipment and to switches or controllers which are used to break
   large currents or are subject to heavy short circuits.
4. *Intrinsically safe circuits* may be used for instruments and telecommunication, including remote control and telemetering by
   light current circuits. With suitable precautions they may be
   used in permanently hazardous situations. Such circuits may
   also be used to supply a few small power consuming devices,
   such as 'safe' hand lamps of low wattage and small motors for
   driving instrument charts if certified for that purpose.

   No flameproof or intrinsically safe or other equipment
   should be used which has not been tested and certified by the
   appropriate authority (see section 11.8.1). It must only be used
   for purposes covered by the certificates and it must be maintained to the standard required by the appropriate specifications. This commonly entails the employment of special skilled
   staff trained and experienced in this type of work.
5. In some situations where the chance of an explosive atmosphere
   occurring and the danger which would result from an ignition
   are *both* small, special reduced precautions may be used.
6. Recently, however, consideration has been given to alternatives,
   largely developed in Germany (see section 12.9 *et seq*).

11.8 Testing flameproof and intrinsically safe equipment

Flameproof, intrinsically safe and other equipment for use in danger
areas is type tested for compliance with the appropriate specifications.
It may be that some equipment will be individually tested if it is for a
special purpose or for an unusually hazardous situation.

The danger of failure depends on a considerable number of ill-
defined parameters and is essentially statistical, i.e. it is impossible to

draw a clear line between 'safe' and 'unsafe' but only to determine the probability of a dangerous outcome. However, with care, this probability can be made extremely small. The basis of such testing and its relevance to flameproof and intrinsically safe equipment is discussed in Chapter 2 'Statistics'. That this approach is satisfactory is clear from the very good record of tested and certified equipment.

11.8.1 Certification

In Great Britain there is at present no legal requirement to use any particular class of electrical apparatus in dangerous situations in industrial premises but 'all conductors and apparatus exposed to inflammable surroundings or explosive atmosphere must be so constructed or protected, and such special precautions taken as may be necessary *adequately* to prevent danger'. General guidance is provided by the Memorandum on the Electricity Regulations (Factory form 928) and CP 1003.

In practice HM Inspectors are unlikely to question the use of apparatus certified to comply with an appropriate British Standard. In some circumstances the national certificates of other European countries or Canada and the USA are accepted; provided that the equipment is properly selected, installed, and adequately maintained.

The first Certificates of Flameproof Construction were issued by Professor Stratham, Professor of Mining at Sheffield University before there were any British Standards. The work was subsequently taken over by the Safety in Mines Research Board (now the Safety in Mines Research Establishment) who still carry out the tests, but from the early 1930's until recently Certificates for Industrial Flameproof apparatus were issued by the Ministry of Mines and those for Intrinsic Safety by HM Chief Inspector of Factories[14b]. Since 1969 all certificates have been issued by the British Approvals Service for Electrical Equipment in Flammable Atmospheres (BASSEEFA). In the Federal German Republic the Standards authority is the Verbund Deutscher Elektrotechniker (VDE) and testing and certification are carried out by the Physikalisch Tecynischen Bundersanstalt (PTB).

11.9 British and German Standards

Because it is the policy of the Commissioners of the European Economic Community, or Common Market, to harmonise national standards so as to facilitate trade between the member countries, efforts are being made by the European Committee for Electrical

Standards, CENELEC, to produce basic specifications acceptable to all members. It is also the wish of the large international companies to have uniform standards throughout their organisations.

In practice this was primarily a matter of reconciling British and German standards which were based on fundamentally different design philosophies. For example, the German practice has been to present ignition occurring by local precautions while the British practice was either to limit the available energy in a circuit (intrinsic safety) for low current circuits, or to assume that ignition would occur sooner or later and contain the explosion and prevent it spreading (flameproof construction). Both methods have had considerable success but they tend to be incompatible. The German method is probably cheaper, but is more difficult to apply without close technical supervision of selection installation and maintenance. It follows that if both systems are to be used in the UK their basic features and differences must be understood.

There will be a transitional stage, which may be prolonged, during which both these systems and a new combined system will be available; and existing equipment complying with the British standard before the days of BASEEFA (see 11.8.1 above) will continue in use for many years because it is efficient and, in the case of flameproof construction, very robust.

11.9.1 Class 'e' increased safety

This was originally a German classification which depends primarily on the elimination of 'open sparking', as at relay and switch contacts or on the commutators or slip rings of motors or generators, and on the control of surface temperatures, partciularly when associated with vapours with very low ignition temperatures. Extra insulation specified creepage distances and enclosures to prevent ingress of damaging materials are also required. For example, squirrel cage motors may need only to have flameproof or otherwise safe terminal boxes but other induction motors will need protection on their slip rings. Small relay contacts can be contained in low volume flameproof or similar enclosures and may be hermetically sealed. The requirements for motor windings present special considerations; fast acting overload and single phasing protection appears to be required, with special desensitising for starting-current rushes, and this may entail complicated arrangements and calculations (Auer and Engle).[12 -]

11.9.2 Class 'N or n' safety

This is a new British classification derived from 'Increased Safety', but appears to be less exacting and therefore it has a more restricted field

of application. Some international companies have produced their own standards but the BSI are preparing specifications. The great difference is that 'Increased Safety' has been used in Germany and elsewhere in situations where flameproof equipment was required in Great Britain. The tendency now is to widen the use of Class 'e' in Great Britain, but I consider that its application will entail detailed specification and attention which only experts can provide and maintain.

11.9.3 Oil immersion

This again is a German standard but in view of the discussion of the danger of sparking or arcing under oil in Chapter 7, I can see nothing in its favour, particularly as only 2-in cover is required (USA, 6-in). See section 10.15.

11.9.4 Sand filling

I am not familiar with any controlled researches on this, but in view of the effective arc quenching properties of the 'sand' in h.r.c. cartridge fuses this may well work. It seems almost certain that dry, uncontaminated sand of sufficient depth would quench vapour explosions and possibly inhibit the ignition of volatile liquids, but any such apparatus should be covered by suitable detailed standard specifications, confirmed by type tests.

11.9.5 Hermetic sealing

This is practicable for small devices with limited energy release under fault conditions. The glass encased mercury tilting relay is an old-established example which unfortunately cannot be considered suitable unprotected in danger areas because of the flexible connections which are necessary and which are liable to break or become detached. An encapsulated magnetically controlled reed relay could be adapted for low currents and voltages. All such devices must be adopted with caution because of their loading and other limitations. Vacuum circuit breakers hold out some promise for power applications.

At best the last two forms of protection would, I think, only be acceptable under Class N in British practice at present and, in my opinion, oil immersion should not be used.

11.9.6 Temperature rise

This has had an important place in German practice. Under the VDE rules equipment is divided into five categories, the permitted surface temperatures being

> G1 above 450°C
> G2 300–450°C
> G3 200–300°C
> G4 135–200°C
> G5 100–175°C

In traditional British practice surface temperature had not been an important consideration, except in a few special situations, because of the use of massive flameproof enclosures, which tend to run cool. The new slightly different BS values have been given previously (see section 11.3.3).

11.10 Selection of equipment

Traditional British practice in industry has been to consider situations as either unsafe or safe; in the former, only intrinsically safe or flameproof equipment was used, while in the latter ordinary industrial designs could be chosen. Applied intelligently this worked quite well. Thus the whole of a process-building replete with piping, pumps, agitators, autoclaves etc was clearly dangerous. So was a building with large storage tanks on the top floor.

Elsewhere when the danger was local, the use of flameproof apparatus could be restricted to a small part of a large site. It is clearly unnecessary to install flameproof lights 20 ft above an installation in the open air where the risk arises from a vapour which is much heavier than air, but this has frequently been done. At ground level, however, the spread of heavy vapours and liquids over the ground and along ducts and trenches must be remembered, (see Chapter 10).

However, the development of large oil, petrochemical and other installations showed it was both unnecessary and prohibitively expensive to use full protection over wide areas and a more rational procedure has been evolved over the years, and practice is still evolving. (The explosion at Flixborough, however, may lead to a reconsideration of the matter and perhaps a tightening up of practice.)

11.10.1 BS 5345: 1976

The first overt step in harmonisation has been taken with the publication of BS 5345 Part I which is based on recent IEC recommendations.

It details 'basic requirements for the selection, installation and maintenance of electrical apparatus for use in potentially explosive atmospheres'. It is to be followed by twelve other parts dealing with specific classes of equipment. These are unlikely to be ready for publication for a considerable time.

In this document the various classes of protection are specified and are set out as in Table 11.3.

Table 11.3
TYPE OF PROTECTION LETTER CODE

Type of protection	Relevant IEC publication	Relevant part of BS 4683	International[5] symbol to accompany 'Ex'	UK symbol to accompany 'Ex'
Flameproof enclosure	IEC 79–1	Part 2	d	d
Increased safety	IEC 79–7	Part 4	e	e
Hermetic sealing	See note 1	See note 1	h	See note 1
Intrinsic safety	IEC 79–00	–	i, ia, ib	i, ia, ib
Type of protection N	See note 3	Part 3	n	N
Oil immersion	IEC 79–6	–	o	–
Pressurisation or continuous dilution	IEC 79–2	–	p	p
Sand filled	IEC 79–5	–	q	–
Special protection	See note 4	See note 4	s	See note 4

Note 1. The symbol 'h' is reserved pending the preparation of an IEC Recommendation for hermetic sealing.
Note 2. Intrinsically safe systems and apparatus may be marked 'i', 'Iia' or 'ib'
Note 3. The symbol 'n' has been reserved internationally pending the preparation of an IEC Recommendation for this type of protection. The symbol 'N' is used in the UK at present.
Note 4. The symbol 's' is used for special apparatus that do not meet the requirements of the standards for other types of protection but that can be shown by test or otherwise to be equally safe
Note 5. For explanation of 'Ex', see page 296.

(The above table from BS 5345 Pt 1 1976 (Table 6) is reproduced by permission of The British Standards Institution of 2 Park Street, London W1A 2BS, from whom completed copies can be obtained).

11.10.2 Hazardous areas, risk points and calamity hazards

Recent practice in the UK has been to class apparatus and installations according to the type and degree of risk in the situations in which they may be used. These were divided into three divisions, more recently called 'zones', in British Standard Code of Practice CP 1003 part 3 according to the risk from flammable gases and vapours that may be present.

Zone (Division) 0.   In which an explosive gas-air mixture is continuously present, or present for long periods.

Zone (Division) 1.   In which an explosive gas-air mixture is likely to occur in normal operation.

Zone (Division) 2.   In which an explosive gas-air mxiture is not likely to occur in normal operation, and if it occurs will exist only for a short time.

These definitions from BS 5345 are simpler than those in CP 1003 part 3.

It is intended that BS 5345 shall eventually replace CP 1003 but it is not clear to what extent it will replace BS 4137 'Guide to the selection of electrical equipment for use in Division 2 areas'.

In Division 0 areas, only intrinsically safe apparatus of the appropriate class was used unless, in exceptional circumstances, pressurisation or similar precautions are taken. Division 1 covers the great majority of situations in which flameproof equipment can be used, while Division 2 corresponds to the 'remotely dangerous' areas of earlier unofficial classifications.

CP 1003 and BS 4137 were not as helpful as could have been wished, because of the wide range of industrial processes and premises and the consequent difficulty of drawing up all-embracing definitions. However, some conditions were included to provide a general understanding of the requirements for safety. BS 4137 while falling short of being a specification, indicates what may be suitable, but this cannot be taken as the last word on this very controversial matter.

The classifications were based on the practice of the large international petroleum and petrochemical industries and the Institution of Petroleum and ICI Codes give valuable advice, and plant manufacturers can help.

Its weakness is that so far no one has been able to give adequate and workable criteria for distinguishing between Zone (or Division) 1 and 2, which are applicable to the wide range of industrial situations, about which opinions vary.

In 1971 Arnaud and Nixon introduced the concept of Risk Points.[13] In large installations there are some areas which clearly fall within Zone 1, for example process buildings where large quantities of inflammable liquids or gases are handled and where the formation of a general or local flammable atmosphere may arise at any time because of leakage, spillage, the failure of an autoclave, or a simple mistake in operation.

There are also large areas in petroleum storage depots, for example, where the probability of leakage is very low except locally, e.g. near

valves, sumps, or immediately below joints in overhead pipes. These are clearly 'Risk Points' where a full flameproof installation may be necessary. The following is a striking example from my own experience. In a large byproduct distillery a joint in an overhead pipeline leaked. It so happened that this formed a puddle near an ordinary (not flameproof) electric motor at ground level. In due course vapour was ignited and the flame flashed along the ground to the benzol house about 20 yards away which was quickly enveloped in flames and the conflagration which followed occupied five fire brigades equipped with foam branch pipes for five hours before it was brought under control. It is to be hoped that this suggestion of Arnaud and Nixon will be followed up in future British Standards or Codes of Practice, as it will greatly assist the man on the job.

In several papers in recent years[14] I have discussed calamity risks (see Chapter 2). These I defined as situations where the probability of ignition may be low or very low, but the possible results are so serious either from damage to plant, property or life, particularly adjacent housing and life, that effective protection must be given, even at considerable expense. The major precautions will not usually be electrical, but suitable siting, installation, design and maintenance of plant and buildings, but there will usually be residual hazards which must be covered by the use of safe electrical equipment. On this matter Chapter 2 and the references to the Flixborough explosion in Chapter 10 should be consulted[15].

Calamity hazards are, however, only extremes examples of a gradation of circumstances and risks varying from cases where, even if there is ignition, no great harm will follow, to those where a major disaster will result with all its political consequences.

It will, I think, be agreed that these varying criteria are all valid. In other words, the distinction between 0 and 1 is easily recognised, but 1 shades off into 2 where some precautions are generally necessary, varying from little more than good standard equipment, well maintained, to risk points where higher or much higher precautions are required, and in determining the risk points and the boundary between 1 and 2 it is necessary to consider not only the probability of a fire or explosion but also the possible consequences.

Such value judgments are essentially subjective but it is to be hoped that agreement can be reached on appropriate levels of protection for most common situations.

11.10.3 Identification

Up to 1969 in the UK flameproof equipment was marked ⟨FLP⟩ together with the *group number* of the gases and vapours for which it was

suitable while Intrinsically Safe equipment was marked to show that it complied with BS 1259 and the *class number* of the gas or vapour. There has, in the past, been no official marking to indicate that equipment is satisfactory to lower standards but very unfortunately an unofficial practice grew up of marking some apparatus as suitable for Division 2 (sometimes merely 'DIVISION 2'). As it closely resembled Group II construction in appearance this can be dangerous as it is likely to be completey unsuitable for situations where Group II would be required.

In future all British equipment will be marked ⟨EX⟩ followed by the appropriate symbol.

BS 5345 gives the following as the appropriate apparatus in the three zones:

Zone 0	Zone 1	Zone 2
Ex ia	Ex d Ex ib	Ex N or n
Ex s — specially	Ex p Ex e	Ex 0
certified for use	Ex s	Ex q
in zone 0	and any suitable	and any suitable
	for zone 0	for zone 1

It will be noted that Ex 0 (oil immersed), is included although it is not at present recognised as a United Kingdom standard (Table 11.5) and it is unlikely to be recognised in future (see above). BASEEFA however tests equipment manufactured for export to countries where oil immersion is recognised.

BS 5345 also recognises equipment which may include more than one type of protection, e.g. a flameproof motor with an increased safety terminal box (ex e or Ex n) and this leads to quite complicated marking, e.g. Ex e, d, 1b, 11c, T3 followed by the BASEEFA certification mark and number. All this is technically sound but it could lead to a considerable duplication of spare parts and units to be stored and identified. A very high standard of technical knowledge will be required and care on the part of electrical fitters, particularly under difficult emergency conditions such as frost, snow and driving rain after dark.

11.10.4 BASEEFA

Developments made in the UK in testing and certifying electrical equipment were described in two articles by the director of BASEEFA, D. E. Fox[16], in 1974 and 1975. BS 4683: *'Electrical Apparatus for Explosive Atmospheres'*, will eventually replace and extend existing specifications, (in particular BS 229 *'Flameproof Equipment'* and BS

1259 *'Intrinsic Safety'*). BS 4683, at present, comprises four parts as follows:

Part 1 Classified permissible surface temperature
Part 2 Construction and testing of flameproof equipment
Part 3 Class N equipment
Part 4 Class e equipment
Other parts will follow.

The new standards, based on the international IEC recommendations, which, however, so far as 2 and 3 are concerned, derive largely from British Standards and practices but Part 2 changes the method of classification.

Because, however, the market for such equipment is international, BASEEFA have carried out testing to the following provisional specifications pending the publishing of new British Standards:

SFA 3002. Petrol metering pumps
     3004. Shunt diode safety barriers (see Chapter 00)
     3006. Battery operated industrial trucks
     3007. Instruments for measuring gas concentration
     [3008. Increased safety, now superseded by BS 4683 Pt. 4]
     3009. Special protection
     3012. Intrinsic safety
     3013. Oil immersed apparatus

It will have been noted earlier, however, that some of these are not acceptable without qualification in the UK, and are intended for exporters.

In the future British requirements will be brought into line with the standards established by the European Committee for Electrotechnical Standards (CENELEC). It is to be hoped that these will not differ greatly from those of the IEC or there may be complications.

Table 11.4 gives a comparison of European (Common Market) Standards and UK Standards. As previously stated the position is still under discussion and further developments are awaited. Some indication of the confusions which remained to be sorted out is shown in Table 11.5 (from R.J. Redding).

To this I would add that although BS 1259 and BS 229 are obsolescent, there is a very great deal of apparatus in existence and use and it will continue in use (and be moved from place to place) for a long time. I would also repeat a warning which I have given elsewhere. Although unification of standards of construction is desirable it may be dangerous to assume that all standards should be acceptable in all countries. *Practice in installation work and maintenance have developed over long periods, and differences in national attitude to technical*

Table 11.4

COMPARISON OF EUROPEAN AND UK STANDARDS

Number	European Standard Title	Symbol	Corresponding UK Standard
EN50.014	General requirements	– –	– –
EN50.015	Oil immersion	Ex O	SAF 3013
EN50.016	Pressurised enclosure	Ex P	–
EN50.017	Sand filled	Ex Q	–
EN50.018	Flameproof enclosure	Ex d	BS 4683 Part 2
EN50.019	Increased safety	Ex e	BS 4183 Part 3
EN50.020	Intrinsically safe apparatus and associated apparatus	Ex i	(BS 1259 and (SAF 3012
EN50.021	Type n apparatus	Ex n	BS 4683 Part 3

Table 11.5

APPROXIMATE* CORRELATION OF GAS HAZARD GROUPS IN
VARIOUS NATIONAL CODES (REDDING)

Test gas or group	IEC group SFA 3012 SFA 3004 BS 4683	BS 1259 Intrinsic safety class	BS 229 Flame-proof class	US national code	German VDE0171 explosion class
Ammonia	–	2a			
Propane	11A	2c	II	D	1
Ethylene	11B	2d	IIIa	C	2
Coal Gas	–	–	IIIb	–	–
Hydrogen	11C	2e	IV	B	3a
Acetylene	[11C in SFA 3012]	2f	IV	A	3b,c,n
Comments	Preferred	Obsolescent		Current	Doubtful

*Because of variations in definitions the comparisons are not exact.

matters may make a device which is safe or reasonably safe in one situation positively dangerous elsewhere.

In conclusion, I have done my best to indicate a path through the maze, as it was at the time I wrote this chapter, but I feel obliged to end with the old mercantile proviso, *e and oe* (errors and omissions excepted).

11.11 Appendix

An indication of the concern felt by the Factory Inspectorate about the indiscriminate use of 'increased safety' may be seen from the following extract from

the Annual Report of HM Chief Inspector of Factories, 1973 (Reproduced by permission of The Controller of Her Majesty's Stationery Office).

'Increased safety' electrical apparatus

In recent years there has been a growing number of enquiries to Electrical Inspectors concerning the suitability for use in potentially hazardous atmosphere areas of imported electrical apparatus of German origin which has been certified by the German standard for increased safety (Ex'c') apparatus. The queries have been dealt with on an individual basis in the light of the conditions obtaining in the premises involved. But the publication by FASEEFA of a Certification Standard for Increased Safety (BASEEFA Standard No. SFA 3008: 1970) makes necessary some remarks of general guidance.

The BASEEFA Increased Safety Standard is based wholly on the International Electrotechnical Commission Recommendation 79–7 which, in turn is derived mainly from the German VDE Standard 0171: 1963 to which PTB have certified apparatus for many years. The basis of the increased safety approach to preventing electrical ignition is the adoption of special measures to ensure as far as possible that no faults which could cause sparking, arcing, or overheating can occur in normally non-sparking electrical apparatus such as squirrel cage induction motors and lighting fittings. These special measures include temperature limitation and minimum requirements for terminals, insulation characteristics, strength of enclosure and degree of protection against ingress of foreign bodies either solid or liquid. This is exactly the approach that is used in the Division 2 non-sparking apparatus method which has been increasingly adopted for 'remotely hazardous' areas in this country over the past ten years, and the only real distinction between the increased safety and the Division 2 apparatus concepts is in the severity of the special measures prescribed.

The similarity of the two concepts does not however mean that the use of increased safety apparatus must be entirely restricted to the remotely hazardous Division 2 areas, for the special measures are, in general, more rigorous. But caution must be exercised in the introduction of increased safety apparatus into the more hazardous Division 1 areas, as, compared to flameproof apparatus, there are a number of factors governing the selection and use of increased safety apparatus to which special attention must be given. These include:

1. The type of enclosure
2. Excess current protection
3. Installation and maintenance of terminal connections

Varying degrees of enclosure protection against ingress of solid foreign bodies and liquids are permitted by the increased safety standards. Where there are bare conductors within the enclosure the degree of protection must be to at least category IP 54 of IEC publication 144 (protection against harmful deposits of dusts and liquid splashing in any direction). But enclosures containing only insulated conductors are permitted to be JP 44 (protection against bodies of thickness greater than 1 mm and liquid splashing in any direction). Further, the Standards include a specific relaxation for electric motors and other rotating plant. These, provided they 'are installed in clean rooms and are regularly supervised by trained personnel', need only have enclosure to IP 20 (protection against contact by fingers, no protection against liquids). It seems possible that, in a Division 1 situation, a ventilated enclosure might not offer a sufficient degree of protection against internal failure resulting from mechanical damage to or chemical attack of

insulation and that, in Division 1 areas, it would be wise to restrict the use of increased safety apparatus to units with enclosures to IP 54 or better. This restriction should not, however, preclude the use of totally-enclosed fan-cooled motors with enclosures to only IP 44 in situations where there is no source of leakage or spillage near to the motor which could create a localised risk point liable to contaminate the air swept over the motor.

The prime difference in the requirements for excess current protection of increased safety as opposed to flameproof apparatus is that account must be taken of transient localised heating of internal conductors. This is essential to ensure, for example, that the rotor bars of a cage induction motor do not exceed the ignition temperature of the gas involved during starting or stalled conditions. The Standard requires that all increased safety apparatus is clearly marked to show all the thermal limitations which must not be exceeded, but the onus is entirely on the user to select the control gear which will meet the increased safety temperature criteria. In this context it is also worth emphasising that although there is naturally no grouping or classification of gases in relation to spark ignition, in the non-sparking increased safety concept all increased safety apparatus is classified in Temperature Classes 1-6 in accordance with IEC 79-8. It is, of course, essential that each unit of increased safety apparatus has the T class number appropriate to the surface ignition temperature of the gas(es) involved at the point of use. Finally, although the Standards require the provision of adequately designed spring-loaded or positive-locking terminals to ensure against loose connections, it cannot be stressed too strongly that safety is ultimately dependent in this respect on good installation and maintenance practice. Experience in this respect of flameproof apparatus has not been entirely reassuring. It is not uncommon for Inspectors to report that they have drawn an occupier's attention to loose bolts on a flameproof joint. But while one loose bolt among many on a flameproof joint would probably indicate, at the worst, a reduction in safety factor, a single poorly-tightened terminal on increased safety apparatus can invalidate the whole 'non-sparking' concept and lead to immediate danger.

References and Bibliography

(*Note.* Probably the best accounts of early researches on flameproof construction are 1 and 2 below)

1. Statham, I. C. F. and Wheeler, R. V., 'Flameproof electrical apparatus for use in coal mines'. *Safety in Mines Research Board Paper No. 60* HMSO (1930)
2. Rainford, H. R. and Statham, I. C. F., 'Flameproof electrical apparatus', *Fuel in Science & Practice* (1931)
3. Ministry of Fuel and Power, 'A review of electrical research and testing with regard to the flameproof and intrinsic safety of electrical apparatus and circuits', HMSO (1943)
4. Whitney, W. B., 'Flameproof certification and its background', *Electrical Journal* (July 1955)
5(a). Phillips, H., 'Flameproof and intrinsic safety and other safeguard in electrical instrument practice', *IEE Conference Proceedings Papers* (April 1962)
5(b). 'Electrical safety in hazardous surroundings', *IEE Conference Proceedings Papers* (March 1971)

5(c). 'Electrical safety in hazardous surroundings', *2nd Conference IEE Proceedings Papers* (Dec 1975)
(The above Conference Proceedings contain a number of important papers by individual contributors)

6(a). Phillips, H., 'A reaction rate theory of flameproof enclosure' (see 5(a) above)

6(b). 'The transmission of an explosion through a gap smaller than the quenching distance, *Combustion & Flame*, 7 No. 2 (June 1963)

7. Cartwright, J., 'The application, certification and testing of electrical equipment for flammable atmospheres', *Electrical Review* (1st March 1968)

8. Redding, R. J., *Intrinsic safety*, McGraw Hill London

9. Slack, S. and Woodhead, D. W., *Proc IEE* 113 (2) (1966)

10. Elterton, G.C., 'Flameproof and intrinsic safety and other safeguards in electrical instrument practice', *IEE Conference Papers* (April, 1962)

11. Eastwood, D. C., 'Application and design of flameproof transformers and switchgear, *IEE Conference Paper* (March 1971). See 5(b) above

12. Auger and Engle, 'Measures of protection of practical experience with Type C explosive-protected electrical machines', *IEE Conference Paper* (March 1971). See 5(b) above.

13. Arnaud, F. C. and Nixon, J., 'Risk point determination as an alternative to hazard area classification', *IEE Conference Paper* (March 1971). See 5(b) above.

14(a). Fordham Cooper, W., 'Electrical equipment for areas of low explosive risk', *Elec. Supervisor* (Dec 1967)

14(b). Fordham Cooper, W., 'Electrical hazards in danger areas, classification of risks and certification of equipment', *IEE Conference Paper* (March 1971). See 5(b) above.

15. *Report on the Flixborough explosion 1975*, HMSO

Chapter 12

Particular Fire Hazards

This chapter describes briefly the application of the principles discussed in Chapters 10 and 11 to some particular situations. This is not an exhaustive study of industrial fire hazards but will indicate the manner in which those principles may be applied in general industry, but the recommendations made must be interpreted in the light of those two chapters.

Two associated problems, namely flame failure protection in gas-fired furnaces and the automatic detection of fires and conditions making fires imminent, together with automatic fire protection are dealt with.

12.1 Installations in danger areas [1]

Engineers concerned with major installations in dangerous areas will be well aware of the difficulties that these present. The following notes are intended primarily for those who are responsible for smaller works or local hazards. Where there is a risk from flammable gases and vapours only safe apparatus of the appropriate type, as indicated in Chapter 11, should be installed. In most situations it is desirable for the equipment to be installed by a contractor who specialises in this type of work.

The first choice would usually be to use metal-sheathed and armoured cable and the cable must be connected to apparatus with approved vapour-proof glands and armour clamps. Trunking should not be used because it may provide a path for the collection and spread of vapour or flammable liquids and a free path for flames. If conduit is used it must be mechanically robust; solid drawn steel conduit or its equivalent is essential. The conduit should have secure vapour tight joints (e.g. screwed connectors with back nuts as would be used for a

supply of gas under pressure) and perfect earth continuity. Where conduit passes from a safe area to a danger area it must be sealed with compound and special fittings are made for this purpose. In an explosion in a garage petrol pump which I investigated, the source of ignition was not the electrical equipment of the pump itself but a fuseboard in a 'safe' area and the flame was propagated along the conduit.

Conduit for intrinsically safe metering or control circuits must also be sealed in the same way. It is extremely important to make certain that any barrier unit (Chapter 10) cannot be accidentally by-passed electrically. Power and intrinsically safe circuits must not be in the same duct or conduit.

Oil-immersed electrical equipment should not be in a danger area even if certified as flameproof, unless it is absolutely essential. Flameproof fuses and control equipment should be excluded so far as possible, since all these items may be under heavy strain in the event of a severe internal fault. For the same reason every effort should be made to keep down short circuit levels; and sensitive earth leakage protection is important.

It is also important that electrical apparatus should not be below or close to joints or valves in pipelines carrying flammable fluids which may leak. Mains-fed portable apparatus should be excluded except for a small class of apparatus specially certified, which is supplied by pliable armoured cables as used in coal mines.

Lighting fittings for *local hazards* (such as a spray booth) may not need to be flameproof so long as they are sited *well above and away from the hazard*, beyond the possible spread of vapour; they should be totally enclosed and robust (e.g. weatherproof designs). For local hazards in large open-air situations with free air-flow, it is also possible to use non-flameproof fittings but these should not be used where there is a possible calamity hazard (Chapters 10 or 11); for example, where a very large amount of vapour may be suddenly released.

When considering lighting fittings, account must be taken of the difference in behaviour of light gases such as hydrogen, methane and ammonia which rise and may be trapped in an 'inverted tank' at ceiling level.

Special care must be taken with naphthalene (or any similar substance) which sublimes, i.e. passes direct from solid to vapour and tends to recondense as a solid inside electrical equipment.

12.1.1 Compo gas pipes

Lead or compo gas pipes which may sometimes be found in workshops or small factories are a serious hazard, as they may be punctured by

quite small earth-leakage currents. When, in an experiment, a short circuit to a compo gas pipe from a circuit protected by a 5A fuse was simulated, in every test, the pipe was punctured and the gas ignited before the fuse blew (Figure 12.1). On one occasion in a workshop current passed from badly earthed conduit to a compo gas pipe and

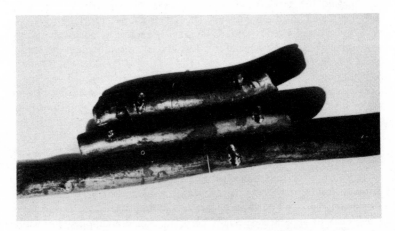

Figure 12.1 A compo gas pipe punctured and the gas ignited by simulated fault from a circuit protected by a 5 A fuse

thence to a water pipe. The short circuit ignited the gas, but the flame also punctured the water pipe and the jet of water put out the fire. This, however, cannot be relied upon.

12.2 Garages and filling stations

Although petroleum refineries, storage sites and pipelines provide some of the most serious fire and explosion hazards, including calamity risks, garages can also be a fire risk. A number of fires have occurred in paint-spraying booths. Administratively, these risks fall under two headings; e.g.

(a) Filling stations, including those on the forecourts of large garages. These are licensed by the local government authorities under the Petroleum Acts.
(b) Motor-vehicle repairing, which comes under the Factories Acts and Regulations.

Table 12.1[1]

TYPICAL ANNUAL FIRES ON PREMISES CONNECTED WITH OIL AND
PETROLEUM INDUSTRIES IN ENGLAND AND WALES OVER A
FIVE-YEAR PERIOD

	Mineral-oil refining	Oil or petrol storage tanks	Oil not in large tanks	Petrol not in large tanks	Oil or petrol pipeline	Garages for petrol-propelled vehicles
Year 1	4	16	24	36	–	392
Year 2	15	10	40	45	5	335
Year 3	16	20	14	24	4	376
Year 4	16	8	120	20	8	400
Year 5	4	20	72	32	8	328

It is impossible to treat the whole of a garage as a danger area, but
the following precautions may be taken:

(a) A sound installation must be provided, particular care being
given to earth continuity.

(b) If pits are used, electrically driven tools should be excluded, and
lighting should be from flameproof floodlights in the walls.

(c) All installations closely associated with petrol pumps and petrol
storage should be flameproof. Portable lights should be intrinsically safe. Dry-battery hand torches are usually inherently safe,
but certified types are available.

(d) Petrol should never be used for cleaning; there are other and
safer liquids. Petrol pumps can be tested in relative safety with
white spirit (see also section 10.16).

12.3 Paint spraying

Paint spraying, which is mostly, but not exclusively, used for cellulose
lacquer and other paint with highly volatile and flammable thinners,
may be carried out in two ways. The paint may be ejected from a hand-
gun using compressed air, or it can be deposited under the control of a
high-voltage electrostatic field from a hand-held or fixed atomiser.
After spraying, tears of excess paint can be removed electrostatically.

There are thus two hazards; i.e. a fire risk in both systems and a
shock hazard with the electrostatic system. This arises not so much
from the sprayer as from the associated high-voltage equipment, in
which it resembles high-frequency plastic heating, and the precautions
for interlocking are much the same (section 13.24).

With compressed-air guns, the danger arises chiefly from heaters,
lighting fittings, extraction-fan motors and small switches. Where these

are within the reach of spray residues, they must be flameproof, and the temperature rise must be limited, since cellulose-lacquer residues may ignite spontaneously at fairly low temperatures. As a rough guide, the surface temperature should be kept well below 100°C for safety. Fan motors should not be inside spray booths or exhaust trunking, since these have been a common cause of fire. Even when they are installed outside, care must be taken to prevent back circulation of air or leakage along the driving spindle of a fan into the motor casing.

Rather surprisingly, electrostatic spraying can be made, in effect, intrinsically safe, but, if an insulated object is sprayed, e.g. a man wearing rubber shoes, he may accumulate sufficient charge to cause an incendiary spark.

In spite of precautions, fires do occur. Friction at the tips of fan blades can be a cause, and automatic detection and extinction are desirable, especially for electrostatic spraying. These remarks apply particularly to cellulose-lacquer sprays, but fires have occurred with less flammable paints e.g. stoving enamel.

Advice on electrical installations for paint spraying will be given by HM Electrical Inspectors.

12.4 Aeroengine and motor-vehicle engine testing

Following serious fires and explosions, a special code of regulations was introduced for aeroengine and accessory testing[2], and *this forms a useful guide for test rigs for motorcar engines*. The regulations deal at length with electrical precautions.

This is a highly specialised risk and the regulations, which should be consulted, go into considerable detail.

12.5 Electrical heat-treatment furnaces

In recent years the use of flammable gases in connection with heat treatment of metals, particularly steel, has rapidly increased, and this inevitably introduces some risk. To avoid the risk of explosions, it is standard practice to scour out the furnaces with gas and burn it off at the far end, before the furnaces are heated up. The persons controlling such operations are usually well aware of the danger of an ignition of a mixture of gas and air and take adequate precautions.

There must, however, be some risk remaining when electric power and gas mains are connected to the same piece of apparatus, owing to the possibility of leakage current passing to earth through the gas mains. For this reason it has been recommended that all metalwork,

including both the conduit or armour sheathing and the gas mains, should be bonded together and effectively earthed locally. At some convenient point not too near the furnace, there should be a break in the gas main with a fire-resisting insulated packing to prevent any leakage current, however small, from passing indirectly to the gas pipes.

12.6 Textile and clothing industries

Up to the present we have dealt with the risks in connection with flammable gases, vapours, dusts and other special substances. However, there are also some risks arising with the processing of materials.

The chief centres of danger in the manufacture of textiles are in the preparation and spinning of the material prior to weaving. These processes are generally accompanied by the evolution of a great deal of fluff which collects on pendant lamps and may easily be set alight by a short-circuit at the cord grip.

Ignition may also take place on open or cleat wiring or at exposed cable and makeshift joints and some very serious fires reported in the past have started in this way.

12.7 Dust explosions

Dust has been the cause of some of the most violent industrial explosions. Metal powder explosions, such as with magnesium and aluminium powders, are liable to be particularly violent, but organic dust explosions, such as cork, are only marginally less destructive.

It will be realised that the concentration of powder necessary, such as 2½ lb of aluminium in a room of 1000 ft^3, represents something like a 'pea-soup' fog, and this is unlikely to persist, except inside such apparatus as grinders, sieves and conveyors. Explosions usually start in such machines, which burst, and the pressure wave so released disturbs accumulated dust on roof trusses and similar ledges, thus producing the necessary cloud as the flame advances. The pressure rise may cause extensive damage, but the velocity of flame propagation may be slow, compared with a gas or vapour explosion. The flame temperature of magnesium and aluminium explosions is very high, and anyone closely involved is unlikely to survive. The mechanism of initiation is often obscure.

An electric arc or the blowing of an 'open' fuse will certainly ignite a dust cloud quite easily, whereas a spark will ignite it rather less easily, so that it is possible to start an explosion inside electrical equipment; but, on the whole, this must be rare. Because of the calamity risk which

may arise, however, adequate precautions must be taken by the limitation of the temperature rise of casings on which dust may collect and by the use of dust-tight equipment.

Flameproof equipment has been used with greased flanges in particularly hazardous situations, but this may be unduly heavy and expensive. BS 3807 sets out a test for dust tightness, and the ERA is able to carry this out. However, there is no constructional standard, so that it would not be impossible for an unduly flimsy casing to pass the test and there is no procedure laid down for certification. The present position is therefore unsatisfactory.

It has been recommended that, where there is a dust-explosion hazard, 0·5A earth-leakage protection and phase-failure protection should be provided. Without expressing any opinions on this point, I feel that a substantially lower figure is advisable when sensitive explosives are at risk.

One particular hazard is the illumination of the inside of grain and similar silos where there have been a number of explosions. The best precaution is to use bulkhead fittings in the walls, with access for servicing from the outside (Figure 12.2).

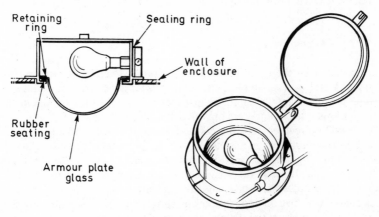

Figure 12.2 Diagram of dust-tight illumination system
For the illumination of silos and bins, the provision of a permanent lighting system using dust-tight electrical equipment is recommended. Facilities should be provided so that the equipment may be maintained from outside the enclosure.
(Safety Health & Welfare Pamphlet 21)

The principal precaution where there is a dust hazard, however, is not an electrical one, but good housekeeping. Key interlock systems such as those that have been developed for high-voltage switchgear, are, however, valuable for ensuring that nobody may enter cubicles where

highly dangerous processes, such as magnesium grinding and sieving, are carried out, until the plant is at rest (see 9.3.2, Chapter 9).

12.8 Furnace ignition and flame failure protection

This is a specialised subject and advice can be obtained from fuel experts and from either the British Gas Council or the local Gas Board.

Danger of an explosion can arise from an incorrect lighting-up procedure or flame failure as a result of either of which a furnace and flues may be filled with an explosive fuel-air mixture which explodes when an attempt at re-ignition is made, or sometimes from glowing deposits in flues etc. Electrical spark ignition is commonly used with flame failure devices to ensure that the pilot primary burner is burning. The placing and adjustment of ignition devices and detectors is a matter for fuel experts, but electrical engineers may be called upon to advise on the electrical controls and circuits, which should be judged in the same manner as other circuits requiring high reliability and failure to safety as described in Chapter 10.

The following are the general requirements as set out in the Gas Council 'Handbook on Industrial Gas Controls' which also contains specific requirements for certain functions and details.

12.8.1 Combustion safeguards

The majority of dangerous situations in gas-heated industrial equipment are associated with the lighting up operation. For this reason a satisfactory flame protection system shall protect against both:
(a) incorrect lighting procedure and
(b) flame failure from any cause.

Some of the devices described in this section i.e. the direct acting thermal expansion and thermoelectric types, are capable in themselves of satisfying these requirements. Other devices are, in effect, flame detectors and an ancillary device should be used in conjunction with them as the safety shut-off valve.

A. *Indispensable characteristics*

Any flame protection system that does not comply with each of the following requirements is unsatisfactory for industrial use.

Every flame protection system shall:
(a) Ensure that the correct lighting procedure, appropriate to the type of burner appliance, is followed.

(b) Entirely prevent gas from being supplied to the main burner until a pilot flame is established

or

Prevent the full gas rate from being passed to the main burner until a flame at a lower rate, subject to an appropriate trial for ignition period, is established.

(c) Be free from any inherent weakness in design which could give rise to failure to danger, provided each component is fitted correctly.

(d) Stop all gas flow to the burners after flame failure and then require manual resetting.

Thus the use of an unprotected pilot, in any installation fitted with a flame protection device, is unsatisfactory for industrial use.

(e) Be actuated only by that part of the flame which will always ignite the main flame, i.e. the flame detector shall not be actuated either by a flame which cannot directly light the main flame or by a flame simulating condition. This is dependent upon correct application as well as on correct design.

(f) Be provided with a 'Safe start check' to prevent energising of the gas valve(s) and, where applicable, any electrical ignition device should a 'flame on' condition be present prior to ignition. (There are a few exceptions to this).

(g) Be mechanically and electrically satisfactory and readily serviced.

B. *Desirable features*

In addition to the above, there are a number of features which are desirable.

Every flame protection system should:

(a) Be protected against adjustment by unauthorised persons as far as is practicable.

(b) Work under all throughput and draught conditions together with all gas characteristic and mixture ratio changes that might normally be expected.

(c) Not be affected in performance by foreign matter.

(d) Operate satisfactorily within an ambient temperature range likely to be encountered in a gas fired appliance.

(e) Tolerate reasonable vibration and shock.

(f) When mains powered, operate satisfactorily with supply voltage variations between −15% and +10% of the normal rating.

(g) Where it incorporates a gas carrying component have an approved capacity within a permitted pressure loss.

To these official recommendations I would add that the general 'philosophy' of Chapter 9, and in particular failure to safety and 'spanning the operation in space and time' should be followed.

12.9 Automatic fire detection and alarms[5]

An incipient fire may be detected by a rise in temperature, a visible flame or glow, by products of combustion, e.g. smoke or gas, or by noise. To ensure that conditions are continuously monitored over long periods some form of instrument or detector is usually necessary.

In all cases the detector operates against fluctuating background conditions. For example, if a very small rise of temperature operates the alarm, it may be set off incorrectly by a general rise of ambient temperature unless some means of discrimination is used. The undisturbed temperature may be very different in winter and summer.

The means of discrimination may be incorporated with the detector or discrimination may be carried out at some central point.

12.9.1 Temperature detection

Probably the oldest established method of automatically detecting and operating on a rise of temperature is the overhead sprinkler in which is, in effect, a circular valve is held closed by a small strut in compression. The strut is in two parts held together by solder. This melts at a selected fixed temperature which must be sufficiently above any likely ambient value to prevent inadvertent operation.

An alternative design uses as the strut a glass capsule containing a liquid which will expand and fracture the glass. The operating temperature may be varied by changing the liquid or the size of a bubble in the liquid; systems of loaded levers have also been used. There have been basically two types of sprinkler system, i.e. the wet pipe which is maintained full of water, and the dry pipe which contains air under pressure and admits water only as the air is released, thus avoiding some problems of corrosion and freezing.

Basically the purpose of a sprinkler system is to apply immediately a douche of water at the seat of the fire, but either type can be made to operate fire alarms or other devices automatically. The dry pipe system can be used solely as a detector; water, foam or CO_2 being applied manually or automatically after some form of cross check. If this were done a lower operating temperature could be adopted, but at the expense of some extra cost and complication, with the necessity of special security precautions.

An alternative application of the same principle is to use a stretched steel wire with fusible links at critical points Figure 12.3. This has been extensively used for atomised water sprays, foam, or carbon dioxide protection in many power stations and sub-stations. The temperature rise is very rapid in transformer and switch-house fires and quick action is imperative, while close temperature discrimination is relatively unimportant. In power stations the system also sets off automatic distant alarms and informs the control room and fire-fighting units of the exact situation of the fire.

In another arrangement an electrical relay is used in a circuit, through which a small current flows continuously, and which incorporates fusible links which melt and break the circuit when the temperature

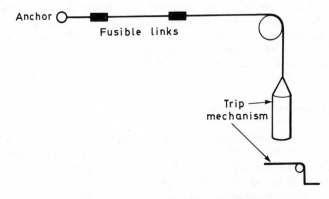

Figure 12.3 Diagram of fusible link mechanism

rises. This has the advantage of failing to safety in the event of a power failure. An alarm working in the opposite manner, i.e. with open-circuited wiring, has insulation whose resistance falls sharply on a rise of temperature and permits current to flow. This principle is used to detect dangerous conditions in aero-engine nacelles, and also to switch off electric blankets well below the danger point when there is an excessive temperature rise.

A somewhat similar result may be achieved by the use of a liquid filled pipe stretching across an area and acting as a giant thermometer. One can ensure better local discrimination if local bulges or containers are connected by small pipes. The length of a single pipe is normally limited to 6 metres.

All these devices suffer from the fact that they operate at a fixed temperature, uninfluenced by other relevant data e.g. ambient temperature.

12.9.2 Discrimination

The temperature will rise more rapidly near the seat of a fire than elsewhere and this may be made the basis of discrimination which may be of two types. That envisaged in BS 3116 is a means of detection which operates at a fixed temperature, e.g. 80°C but responds more quickly to a rapid rise. This may often be achieved by adjusting the thermal capacity of the detector element in relation to that of some heat sink, which may be the metal of its support.

Alternatively, discrimination may be arranged to take account of ambient conditions as well, so that the detector acts more quickly and at a lower temperature for a sudden rise and depends on the rise above the ambient conditions rather than on the absolute temperature. This will generally permit operation at a lower temperature without risking false alarms and may be valuable in hazardous situations. If a closed vessel or pipe is fitted with a pressure sensitive device and a small leak to the outside air at a fairly remote point, its internal pressure will normally follow that outside, but there will be an appreciable delay in reaching equilibrium if the vessel is suddenly heated either by convection or radiation and the internal pressure rises and operates the alarm. This should be backed up by a device operating at a fixed, maximum permissible sustained temperature.

Detection of temperature changes may be made electrically in a number of ways, of which the simplest is probably a bi-metallic thermostat in which contacts are closed or opened by the unequal expansion of two elements one of which may be screened from the source of heat. If both have the same thermal characteristics the device will operate during changes in temperature, while if they have different characteristics the device may be designed to operate more quickly when there is a rapid change but also to operate after an interval on a steady rise. Probably the simplest arrangement is a concentric rod and cylinder, Figure 12.4. This principle could also be used to control the flow of liquid in a pipe, on the lines of the thermostat in a gas oven or boiler, and thus give a remote indication, but I have not met this in practice.

Any of the established methods of remote electrical temperature measurement or indication may be used for detecting fire. If, for example, a resistance thermometer element forms one arm of a wheatstone bridge and a remote or screened element is used as the reference arm, then the instrument will indicate a rise of temperature above ambient. It is not difficult to discriminate automatically between instrumentation faults and dangerous conditions, so that failure to safety does not introduce an undue risk of false alarms. There is virtually no limit to the degree of sophistication which may be applied

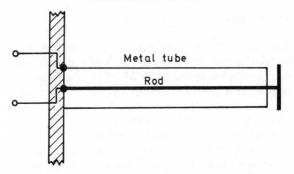

Figure 12.4 Bimetallic concentric tube and rod. If the temperature rises suddenly the tube expands more quickly than the rod which is screened from direct heat

to the automatic interpretation of the reading – if it can be justified financially.

12.9.3 Light and radiation

In some critical situations, i.e. explosive factories, it may be worth while using expensive means of detecting the first small flicker of flame and instantly dousing it with a large volume of water, particularly if the point of potential danger is fairly accurately known. The difficulty is that except in totally enclosed buildings with no windows the background illumination, varying from near zero to bright sunlight, makes discrimination extremely difficult.

Therefore, although a very small light is easily detected by a photoelectric cell, it may be better to use one of the devices sensitive only to infra-red radiation. Alternatively it may be possible to use a bridge circuit to balance the light from the danger point against background illumination at some other point. The instrumentation, however, tends to become complicated and the adjustment critical.

Infra-red radiation is also more easily detected than visible light amid smoke and dust or vapour clouds which is an added advantage, and portable infra red detectors may help fire-fighters to find the seat of a fire amid smoke – if the smoke is not too dense.

12.9.4 Products of combustion: smoke

Smoke is easily observed and is very insidious, easily passing through packed material. It is not surprising, therefore, that it has for a long

time been used to detect incipient outbreaks in sealed cargo holds on ships and similar situations. Small pipes from selected points lead to a closed box with a glass front into which air is sucked and any smoke is easily seen against suitable illumination and background.

To detect smoke automatically is more difficult since no optical instrument has the range of sensitivity and discrimination of the human eye, but successful applications of photo-electric cells have been made both to the smoke boxes described above and by monitoring light beams in the space above cargo, etc. Dust on the lenses can be a serious nuisance, but dust collects on surfaces colder than their surroundings and tends to avoid surfaces which are warmer as everyone knows who has hot water radiators for central heating (this is a result of the 'Brownian movement').

A different approach has met with some success. If the air in a sampling chamber is ionized, e.g. by a small amount of radioactive material, it will become slightly conductive, but smoke particles quickly gather up the ions and greatly reduce the conductivity and therefore quite small amounts of smoke in the air sample can be detected. This device is essentially electrical or electronic and may be used in conjunction with the standard telecommunication and control equipment discussed below.

It has been stated that the molecules of some gases and vapours also 'accept' free electrons and small ions and by increasing their mass and reducing their mobility will have the same effect.

12.9.5 Products of combustion: gases and vapours

The most obvious gaseous products of combustion are carbon dioxide, carbon monoxide and steam and for these any continuous means of detection and measurement can be employed. The detector should preferably be of the type which monitors a stream of exhausted air flowing through the instruments. Modern automatic micro-analysis is a rapidly developing art and it is not possible to discuss all the methods in a short space. They are, however, usually costly and can only be justified in special circumstances.

Quite small changes, e.g. one or two per cent in the composition of air, may be detected by physical means such as testing thermal diffusivity and absorption of light. If cool air is passed over a heated resistance wire forming one arm of a wheatstone bridge the wire will be cooled and its resistance changed by an amount depending on the thermal properties of the gases. This can be detected by comparing it with the resistance of a similar wire forming the reference arm of the bridge and immersed in 'standard' air. This would not be effective unless the air

stream had been cooled. Changes in spectral absorption can be measured by passing light through two parallel tubes, one containing standard air and one containing the sample to be tested.

The second instrument is sensitive to uncombined oxygen which will be depleted and to some other gases but an alternative instrument depends on the fact that, rather surprisingly, uncombined oxygen can be detected and measured in a magnetic field.

So far as the writer is aware, such tests have not been applied to the detection of combusion but they would repay investigation for particularly hazardous or valuable situations. Products of thermal decomposition or incipient combustion such as aldehydes, ketones and peroxides could, in principle, be detected by suitable instruments, but the equipment is likely to be expensive.

12.9.6 Pressure-wave detector

When an explosion occurs a pressure wave travels out ahead of the flame, and this has been used to initiate protection. The velocity may vary from a moderately slow displacement 'puff' in the early stages of a dust explosion to a detonation wave travelling at a speed exceeding that of sound (approx. 3350 m (11 000 ft) per second). It has been possible to detect this and, at least for the slower initial stages, to trigger off the

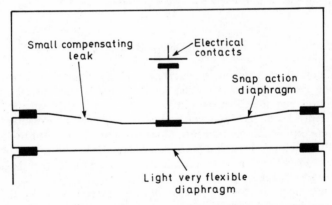

Figure 12.5 Quick-acting pressure detector capable of operating on advancing pressure wave ahead of flame

release of an extinguishing substance and prevent the spread of the explosion. One method is to actuate a pair of electrical contacts directly from a very flexible diaphragm. This can be made stable and also to operate more rapidly for a sudden rise and take account of

ambient pressure changes and temperature by ventilating the space behind the diaphragm via a small hole, as described above for temperature detectors, Figure 12.5.

Such a system must be very sensitive and have the minimum of inertia if it is to act sufficiently quickly, and the optimum positioning of the detector may be important.

12.9.7 Monitoring danger zones

All fire detection systems are in effect monitoring systems, but so far only those which aim to detect things which may soon burst into flame have been considered. Where the risk is high or the possible damage is great, however, it may be reasonable to forestall an outbreak by detecting hazardous conditions before a fire can occur.

Possibly the oldest widely used device for this purpose was the Davy miner's lamp. When 'fire damp' (methane) is present the character of the flame changes and gives a warning of danger, while the lamp is so constructed that it cannot itself be a source of ignition.

Modern equivalents of this are the various types of explosimeter. They are generally designed to make spot checks on the percentage of flammable gas in a sample of air, but in principle the design could be used to monitor a continuous flow of air through the instrument. The most common construction is for a platinum wire to be heated in the gas to be examined as described above, but instead of being cooled it is heated by catalytic surface combustion on the wire and its electrical resistance is changed in the opposite sense to that of the previous application.

Here again any one of many methods of detecting the presence of foreign gases and vapours or changes of concentration of those normally present may be employed, but the instrument needs to be calibrated for the particular gas or gases which may be found. If it is desired only to indicate a change and not to estimate danger numerically, the problem is easier.

Such methods may be used to detect the escape of town gas, hydrogen, oil refinery products such as butane, solvents such as alcohol, acetone, etc, or refrigerants such as ammonia. They could also be used to detect products of slow oxidation such as occurs on the surface of stored coal and other substances which oxidize slowly but may heat up and finally inflame. In such cases, however, it would often be easier to measure increase in temperature.

A slow rise in temperature may also indicate incipient danger to electrical machines or in cable tunnels at power stations and this is

often recorded for other reasons, e.g. because of the risk of insulation failure.

In special cases other factors may be monitored, e.g. insulation resistance and thus the whole wide field of electrical protective gear comes in one sense under this heading.

12.9.8 Operational security

The first requirement is that any alarm or instrument must not itself be a source of danger. This applies particularly to electrical systems — where flammable gases, vapours or liquids may be present — but fortunately a fairly complete technology has been worked out under the headings of Flame-proof, Intrinsically Safe and Dust-tight equipment. This is covered by a number of British Standards which are virtually mandatory (see Chapter 11).

The next requirement is that the detection and alarm system shall not itself be an early casualty in the event of fire and it should fail to safety. (It is virtually impossible to make it completely proof against the effects of a very severe fire when nearly everything else is destroyed.)

The system should also be proof against fortuitous failures. In the first place there must be a secure source of electrical supply and the most satisfactory is a floating battery-operated system with an automatic (trickle) charger and an alarm to indicate when and if the latter becomes ineffective. The source of supply and the control or report centre should also be housed in a building or cubicle which is virtually proof against fire and other hazards.

Any advanced system will no doubt depend on telecommunication apparatus and techniques and the standards adopted in railway signalling merit special study (see Chapter 9).

The reader may also consult CP 1019:1972. This deals with the installation and servicing of electrical fire alarm systems in buildings and covers all systems from the simplest installation to sample installations which can include automatic detectors, control and indicating equipment.

12.9.9 Telephone and loudspeaker

There is a temptation to use the works telephone or loudspeaker system in association with fire detection and alarms and this needs special consideration. The general aspects of circuit protection and freedom from electrical interference are discussed in the CP 1019 mentioned above.

The type of wiring often used in telephone systems has no pretension to fire resistance and would not be acceptable. Apart from this, works broadcasting systems are not as a rule designed to fail to safety; for example, the writer witnessed the test of a fire warning to be transmitted over loudspeakers, and the man who was to give the alarm became excited and shouted at the microphone which overloaded the amplifier, so that a fuse blew and shut down the equipment — and no warning was broadcast.

When alarms and messages are to be transmitted via a manual or automatic telephone exchange they may be blocked by other users, and even where separate telephone circuits are in operation they may not receive immediate attention. Thus, if a warning is transmitted to a power station control room, even if a special telephone receiver is provided, it may well have to take its turn among other incoming calls. The control engineer may have half-a-dozen telephones on his desk and when there is a fire there is almost sure to be other trouble and he may not pick up the 'fire' receiver first, and it is not necessarily the most important, even in the interest of safety.

Such applications must, therefore, be very carefully assessed.

Acknowledgements

The first part of this chapter is based on the paper on Industrial Fire Hazards by my colleague the late F.H. Mann and myself[3], and my review paper on Electrical Safety in Industry[4]. Section 12.8 on flame gratefully acknowledge. Section 12.9 on automatic fire detection is based on an article from *Industrial Systems and Equipment*[5] which has been updated to conform to present-day practice.

References 6–14 list publications which the reader may find useful as a source of additional information on this important subject.

References

1. Emerson, S.J. *Electrical Hazards in Industry, with particular reference to Petroleum Spirit Installations*, Inst. of Weights & Measures, Admin. Conf. publ. 1959.

2. *Aircraft Engines: Factories (Testing of Aircraft Engines and Accessories) Special Regulations* 1952, S1 1689, 1952, HMSO. Their importance is reduced with the development of jet engines, but they remain a valuable code of practice for aero and car engines, and in fact for any situation where large amounts of 'loose' flammable liquids are necessarily handled or used.

3. Fordham Cooper, W. and Mann, F.H., 'Industrial fire risks', *Journal IEE.* **91** Pt 1, (July 1944)

4. Fordham Cooper, W., 'Electrical Safety in Industry', *IEE Review* **1117**, (Aug 1970)

5. 'Automatic Fire Detection Systems', *Industrial Systems & Equipment* (June 1967)
6. *Rules of the Fire Offices Committee for automatic fire alarm installations* (deals with requirements of insurers) Fire Offices Committee
7. CP 1019: 1972 'Installation and servicing of electrical fire alarm systems', BSI
8. BS 3116 Part 1: 1970. Part 2: 1973 'Automatic fire alarm systems in buildings', BSI
9. 'Fire detection devices' *Jour. IEE* 88 Pt 1 (Feb 1941)
10. 'CP 1003. Part 2: 1966. Part 3: 1967. 'Installation and maintenance of flameproof and intrinsically safe electrical equipment (Note CP 1003 Part 1 has been replaced by BS 5345: Part 1: 1976) BSI
11. BS 5345: Part 1: 1976. Code of practice for the selection, installation and maintenance of electrical apparatus for use in potentially explosive atmospheres. Part 1 Basic requriments for all parts of the code. BSI
12. *Fire protection in factory buildings*, Factory Building Study No. 9 D.S.I.R. HMSO
13. *Fire fighting in factories*. Safety, Health & Welfare Booklet No. 10, HMSO
14. Wheeler, L.S., 'Automatic Fire Detection & Prevention', *Jour. of the Junior Institution of Engineers*, (Jan. 1967)

Chapter 13

Everyday Applications

13.1 Introduction

This chapter deals with the subjects on which I had most frequently to advise in my work as one of HM Electrical Inspectors. In relation to particular apparatus and situations, it necessarily covers matters discussed in more detail in previous chapters. I hope that it will therefore clarify the application of general principles to individual cases. To help readers who wish to use this book for reference, cross references to other chapters or sections are given in the outside margin (i.e. Ch 7 or s 7.1).

The Memorandum to the Electricity Regulations (Factory Form 928), published by HMSO and the current IEE Wiring Regulations for the Electrical Equipment in Buildings should also be consulted.

Some of the advice given in this chapter applies to equipment which is or may become obsolescent, but this equipment may still be in use if it is adequate for its purpose. The treatment is not all embracing and there may be situations for which the advice given would be inappropriate.

It is assumed throughout that the common alternating current supply voltage, i.e. 50 Hz, 230/400 V is used unless otherwise stated or implied.

The following is the UK statutory classification of voltages measured at the point of use in the case of low and medium pressure and use or supply for higher pressures:

Low	0 to 250
Medium	over 250 to 650
High	over 650 to 3000
Extra high	over 3000

Extra low and very low are not defined legally but, in practice, are usually taken as meaning about 12 or 25 V a.c. which are very unlikely to cause serious shock accidents.

13.2 Industrial substations

Provision must be made for isolating h.v. circuit breakers, etc for servicing; in small sub-stations, circuit breakers with 'wing isolators' may be suitable. Because of the high fault levels under modern conditions final isolation in larger sub-stations is best effected at circuit breakers with draw out or drop down isolation. Exposed ports or

ss. 7.2,
7.3
Ch. 8

Figure 13.1 A simulated short circuit

spouts should be automatically covered by shutters which are padlocked while work is in progress, but arrangements are necessary to allow the cable spouts to be exposed for testing or earthing while the bus-bar spouts remain locked, unless special, built-in provision is made for this purpose. While any work is in progress, high voltage equipment which could possibly become alive, because of a fault, mistake or by induction must be solidly earthed. Earthing of cables is best carried out

through a circuit breaker. In many designs provision is made for attaching earthing plugs to a withdrawn circuit breaker, but this may entail some danger if they are not used carefully.

When flexible conductors ('jumpers') are used for earthing they must be adequate to carry full short circuit current until they are cleared. This can be related to $[I^2 t]$ per square inch of copper, see Chapter 7. The Chief Engineer of one of the Electricity Boards, in an address to the Institution of Electrical Engineers, recommended that fault levels should be limited to 13 000 A (corresponding to 250 MVA at 11 kV). Figure 13.1 shows the result of a simulated fault on an inadequate earth jumper.

Heating is desirable to prevent condensation in cold weather. 'Heating cables' or heaters should be close to the switchgear or other vulnerable points, e.g. under switchgear spouts, since condensation migrates from warm to cold surfaces.

Safe access, which includes exit, is important. A substation should not include blind alleys or spaces where someone may be cut off by an explosion or fire. Passageways with an exit at each end are best.

Very heavy medium-voltage short circuits may occur electrically close to large transformers; for that reason work near, or on, medium voltage switchboards must be carried out with great care. Open switch and fuseboards with exposed live conductors should not be used.

The most serious danger, however, arises from fires or explosions associated with switch or transformer failures (see Chapter 10). To prevent the danger of extensive spread of fire, indoor switchgear and transformer units should be sectionalised in groups of moderate capacity. This can best be effected by segregation within separate buildings where this course is warranted by the size and importance of the plant. Physical separation in the form of fire-resisting barriers between transformers or sections of switchgear can, however, be an effective alternative in circumstances where such dispersion is impracticable.

The provision of drainage sumps filled with clean washed pebbles or ballast is recommended for the absorption of oil which may leak or be released under fault conditions from transformers or switchgear; this restricts the spread of fire should ignition occur. Such ballast is best graded with the finer aggregate at the top so that, if alight, the oil flames are extinguished by the cooling effect as the oil falls through to the larger interstices below. Where sumps cannot be used, low walls or curbs built around the transformer provide a useful compromise. *ss.10.15 10.16*

Where appropriate, cables may be led out of the switch or transformer enclosure overhead rather than in trenches, which often form channels through which fire can travel, and bunching of cables should be reduced to a minimum. Trenches should be stopped by sand *ss.10.12*

barriers where passing beneath walls, and protection for the cables should be provided by a filling of sand or pebbles. Cables should also be supported where they pass through floors, and the apertures should be sealed. Bitumen serving should be removed from metal-sheathed cables unless it is essential for the prevention of corrosion, and the cables should be protected by fire-resisting covering between the floor and the apparatus to which they are connected.

In substations or transformer chambers situated below blocks of buildings where the fire risk is considerable or in which many people may be employed, more comprehensive precautions are necessary both as regards fire prevention and extinguishing facilities. Only brief reference can be made here to such measures which, it is suggested, should include the following:

(a) The structure and, in particular, any party walls or partitions and doors should be of very substantial construction and capable of resisting the effects of fire for a considerable period.

(b) Independent ventilation of the chamber and the exclusion therefrom of any trunking which is in communication with other parts of the building.

(c) Means to indicate undue rise of temperature in the chamber, and means, remotely situated, for disconnecting the high-voltage supply.

(d) Fixed fire-extinguishing equipment, automatically operated.

(e) Other considerations suggest themselves, for instance the danger arising from the sudden extinction of the lights in a crowded building, due to an electrical failure or fire, implies the desirability of installing an emergency supply, if only of limited capacity.

(f) The possibility of an oil-immersed circuit breaker exploding violently must be considered.

Considerations of this nature are naturally of greatest consequence where oil-filled electrical apparatus is concerned. Such conditions invite attention to the advantages of oil-less equipment.

The development of air-break switchgear for a wide range of voltages is a contribution in combating the fire hazard in confined situations and, similarly, the risk of an oil fire from a transformer failure can be eliminated by resort to air cooling. The use of non-flammable coolants for transformers does not yet appear to have been widely adopted.

13.3 Cables

Cables must be protected from damage unless protected by tough metal sheaths, armour conduit or trunking. No exposed lengths of cable

should be allowed between the ends of conduit and the entry to apparatus. Special care is necessary where cable runs are within reach of workers or liable to be struck by crane hooks, suspended loads, ladders, or fork-lift trucks, etc.

Temporary cable runs must be protected. This may be done for short periods by running them through *tough* garden hose. All temporary work should be dismantled as soon as possible, before it becomes an accepted part of the scenery and forgotten. Twisted lighting flex should never be used for temporary installations. Tough rubber or p.v.c. sheathed flex can be used only for short periods and when protected against accidental damage.

Cleat wiring is only suitable for corrosive situations where metal conduit or trunking cannot be used, but with the development of tough p.v.c. or similar conduit, and fittings this is largely obsolete.   *s10.12*

Pendant lights and similar fittings should have only *circular-sheathed* plastic or rubber flexibles which reduces damage and fire risk. All cord grips must be properly adjusted. Joints must be made by proper   *ss10.12* connectors suitably located and protected, taped twisted joints should   *10.13* never be used.

Ducts and tunnels used for cable assemblies present a serious risk in the event of faults leading to arcing, fire, or heavy earth return currents. In such circumstances p.v.c., for example, may burn although under other conditions it is fairly fire resistant; corrosive or poisonous fumes may be produced when burning. Tunnels are also liable to be filled with dense clouds of black smoke, and ventilation may well cause a fire to burn more fiercely. For these reasons sensitive detectors for over-heating, smoke and/or fumes should be considered in association with automatic fire protection, e.g. atomised water spray or flooding with   *s.12.9* nitrogen or CO_2. Chlorinated hydro-carbons and similar extinguishers should not be used in enclosed spaces.

Cables for different essential purposes should not run in the same ducts or close together in tunnels. For example, a power short circuit should not endanger emergency lighting, control, or important tele-communication channels. Consideration must also be given to the possible damage to control and telecommunication cables by return earth fault current flowing along the sheaths, or the very high earth   *s.8.2.6* voltage gradients which may exist during major power faults. Major power stations have been shut down by such events on more than one occasion.

Very severe through faults may damage cables in two ways; they may be overheated and catch fire and large lead-sheathed oil impreg-nated paper insulated cables may be burst by boiling the impregnating oil. The insulation of a cable can act to some extent as a heat sink, therefore the cable ends or tails may be much hotter than the rest of

the cable and in extreme cases the copper may evaporate, leaving an empty tube. On other occasions the electromagnetic forces on the conductors may force them apart and burst the insulation and covering. I have observed all these results while investigating failures, but on one occasion what appeared to be an internal 11 kV cable fault to earth below a circuit breaker, which melted the cable sheath, proved to be an arc from the switchgear frame to the sheath caused by inadequate bonding. The cable was undamaged internally. In this case the facts were only made clear by dissecting the cable. If this had not been done, money would have been wasted on useless safety precautions and the real causes probably left uncured.

13.4 Cable joints and dividing boxes

Some years ago in the north of England, there was an unusual number of failures of buried-cable joints, some of which were explosive and potentially dangerous, which appear to have had two principal causes: ineffective filling, leaving voids which lead to ionisation, and/or the ingress of moisture, and the uprating of old 6·6 kV cables to 11 kV.

There has also been a series of explosive failures of dividing boxes on motors, switchgear, transformers and busbar trunking. These have occurred at medium and high voltage and can do considerable damage. In one failure, part of the casing of a dividing box hit a man on the head and killed him. Some of these were again a result of bad filling leaving voids, and the late D. A. Picken suggested Xray or γray examination of existing equipment. Other failures seem to have been breakdowns caused by voltage spikes. Several papers and alternative proposals for preventing these troubles have been published[1 and 2], the best of which appears to be the use of more complete terminal insulation and phase segregation or effective insulating barriers between phases, and more consideration should be given to the prevention of voltage spikes generated by relay coils and h.r.c. fuse operation on medium-pressure equipment and also possibly at 3000 or 6000 V.

s6.15
ss 7.7; 7.9

Everyone who has investigated a considerable number of failures of large cartridge fuseboards and boxes has probably met examples of flashover which are difficult to understand unless they were due to voltage spikes caused by current chopping arising from some unusual combination of circumstances.

Bolted cable joints may be loosened as the result of differential expansion on heating and this may lead to further overheating and sparking or arcing and finally damage or failure. It is particularly liable to happen when different metals form the joint (including the bolt), see Reference 3.

13.5 Cable loading (see Chapter 7)

The safe loading of medium and low voltage cables in industry is in practice determined by the tables published with the IEE Regulations. Attention should, however, be drawn to two points. Momentary current rushes on switching on filament lamp lamps, transformers, etc are unlikely to damage cables since the withstand $[I^2 t]$ will usually be adequate but it may be greater than the pre-arcing $[I^2 t]$ of fuses and specially designed or selected cartridge fuses may be required.

On the other hand, third harmonic currents caused by fluorescent lighting loads, and to some extent by transformer flux saturation can cause excessive heating because the $[I^2 R]$ losses of the fundamental and the individual harmonics are additive. ERA Report 5086 recommends a derating to 85% of the normal for all multi-core cables and single-core metal sheathed cable up to 0.5in^2 and for larger cables to 75% and, because the third harmonic currents will return via the neutral, which should be as large as the phase conductors.

13.6 Earthing (see also Chapter 8)

All exposed metalwork which could become alive must be earthed. The only exception is in earth free areas as used for testing by skilled staff, and certified double insulated portable appliances. *s 13.23 Ch. 4*

The joints in earth continuity conductors must be properly made and mechanically reliable, so that they cannot inadvertently come undone or break. Lengths of flexible metallic tubing at the entry to motors, etc. mounted on slide rails are not efficient earthing conductors since they often become detached or loose at the ends or fracture. An independent earth wire or other connection is necessary.

The steel frame of a large building may be a good earth, but it should be tested. Additional earthing to special earth plates (electrodes) is often advisable, but water pipes are becoming unreliable because of the increasing use of plastics.

In many situations, particularly in machine shops with overhead trunking, reliable earthing is difficult; it is useful to lay a substantial copper strap say 25–38 mm by 3 mm firmly pegged down on the floor and bonded to all machines. Strength is just as important as continuity, so wherever possible have two or more earth paths in parallel.

Earth connections must, of course, carry the maximum possible earth fault current in safety. This is very important where heavy short circuits are possible. It is also necessary for the earth loop impedance to be sufficiently low to pass enough current to blow the fuse or operate *s 8.3.2*

overload protection unless earth leakage protection is provided. Assuming a fusing factor of 2 and a factor of safety of 2 the *maximum permissible* loop impedance is $Z_m = V/4I$ ohms where I is the fuse rating and V is the voltage to earth. (The IEE regulations allow somewhat lower values but it is best to be on the safe side and this has been the Factory Inspectorate recommendation.) The IEE Wiring Regulations go into a good deal of detail on earth continuity and should be carefully studied.

13.7 Low and medium voltage isolation

s 9.3

Means of local isolation from the supply must be provided for *all* electrical equipment except at very low voltages, so that maintenance and adjustments may be carried out in safety, or the supply cut off in emergency. For this reason a load-breaking isolating switch is preferable to an isolator or links. This should be placed as close as possible to any motor starter or other controller so that when making adjustments to these, an electrician has the isolating switch within reach.

As they can be used by persons without electrical skill or training, isolating switches are also convenient for use by operators, millwrights, toolsetters, fitters and others who have to work on machinery.

If the isolating switch itself needs attention, which should be rare, isolation must be carried out at the other end of the incoming cable. The construction of isolating switches, or switches and fuses is discussed in section 9.3.

s 13.17

Isolating switches for overhead crane trolley wires must be provided at working floor level for use in emergency. It may also be necessary to have isolating switches at a higher level for the use of maintenance staff. Similarly where a motor controller is on one floor and the motor on another it may be desirable to have a switch adjacent to the motor as well as covering the controller. Power station induced draught fans for boilers are an example.

s 9.3.2

Because isolating switches are used for making electrical and mechanical plant safe for people to work on they should be capable of being locked in the open position. Some switches have built-in locks, in other cases the handles can be padlocked.

Figure 13.2 illustrates the location of isolating switches on a simple distribution system. *Small* enclosed or cartridge fuses in insulating holders, where the rupturing duty is not high, if properly constructed (see section 7.6), may not need to be provided with isolators if they can be withdrawn with safety. There is a risk of burns if semi-enclosed fuses are re-inserted in a faulty circuit but this is obviated by the use of correctly selected cartridge fuses.

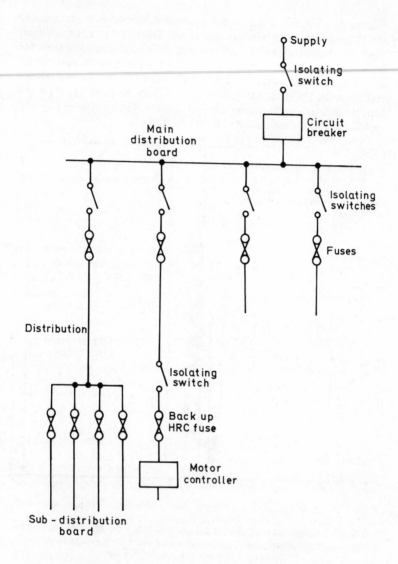

Figure 13.2 Layout of simple distribution system

13.8 Switch- fuse- and control boards

High and medium voltage switch and/or fuseboards with exposed live metal, once common, are rare nowadays. There is no excuse for their existence except in a few special cases, such as contactor panels for large motors, e.g. for rolling mills, or the control of processes. Where used they must be in enclosed places to which only adequately trained and authorised persons have access. There must be good lighting and a firm insulating floor and adequate passageways. The clearances required by UK regulations are indicated in Figure 13.3.

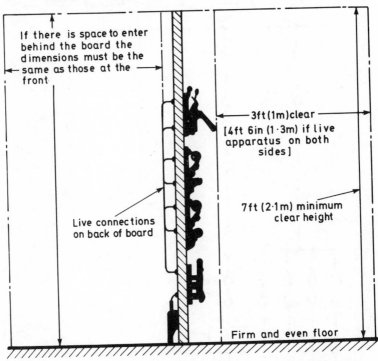

Figure 13.3 Low and medium voltage switchboard passageways (statutory minimum dimensions).
For higher voltages larger clearances are laid down (see the Memorandum on the Electricity Regulations) but open switchboards are strongly deprecated unless there are exception reasons for their use

Special considerations arise for some apparatus with high voltage contactors, thermionic valves, etc. Examples are the control of high frequency metallurgical furnaces and radio frequency glue setting and

plastic heating, high voltage motors etc. It is essential for maintenance electricians to watch contactors in operation but they cannot do so if prevented by door interlocks; for this reason observation windows must be provided. The same applies to the examination of large thermionic rectifying and other valves. If output falls it may be caused by the failure of one of a bank of valves and the easiest way of locating the faulty one is to watch them in operation. Fatalities have occurred when interlocks were defeated for this purpose – conductors at up to 11 000 V being exposed.

A further difficulty arises in checking the operating coil circuits, as it is often desirable to have a very compact layout. The recommended procedure is to have full interlocking between the door lock and the main power supply (preferably by direct mechanical means, see Chapter 9) and to supply control circuits at a low and relatively safe voltage such as 50 V (voltages below 50 are apt to give trouble on low current circuits from dust and dirt on contacts). (Interlocking is discussed in Chapter 9.) Care must be taken to ensure that after a door is opened *s 9.9* the switch cannot be re-closed until after the door has been re-closed. This is often more difficult to achieve with certainty than might be expected.

With such interlocking care must be exercised in eliminating sneak circuits. One electrician was electrocuted because a wattmeter with *s 9.7* 400 V back terminals was mounted on a low voltage control panel. It had no direct electrical connection to the control circuits, but was for *s 9.8* reference in the operation of the equipment.

These principles can with advantage be applied to the control of medium voltage plant as well.

13.9 Fuses (see also Chapter 7)

Fuse assemblies should not have exposed live metal which can be inadvertently touched and they must not scatter hot metal or gases when they operate. This is avoided by using cartridge fuses in well-designed holders.

Metal-clad switch and fuseboxes are often designed so that they cannot be opened until the fuses are disconnected from the supply. *ss 9.3.3;* This is particularly true of large fuse cartridges which are bolted into *9.3.4* position to ensure good contact. With this construction the incoming connection to the switch must be effectively shrouded.

Fuses must have adequate rupturing capacity and have an operating or cut-off current sufficiently low to protect outgoing cables and *ss 7.4* apparatus (see section 7.3). They must also operate on earth faults *to 7.15* earth leakage circuit breakers or relays are not provided.

The fuse ways should be clearly labelled, or a fixed diagram provided to enable the fuses of particular circuits to be quickly identified, and the correct fuse size should be stated. This *may be lower* than that required to protect the cables and vary according to the apparatus connected, which may change from time to time. Fuse sizes should only be changed by persons with adequate technical knowledge and any changes should be recorded.

The selection of fuses is dealt with fully in Chapter 7.

13.10 Lights and lighting fittings

Metal lighting fittings, including lampholders, whether fixed, pendant, or portable, must be earthed unless they are supplied at a very low voltage, e.g. 12 V. Fittings supported by a chain are not effectively earthed unless there is also an earth wire. A short steel chain can have a resistance not far short of 100 ohms and contact is completely lost if the fitting is lifted slightly (this caused at least one fatality). This may not be absolutely essential in earth free areas but to omit the earth wire may be inadvisable since the reason may be misunderstood. The IEE Regulations call for provision for earthing all metal lighting fittings but grant an exemption from earthing where they are over a non-conducting floor and cannot be touched by a person standing on or within reach of earthed metal. This is not likely to arise in industrial situations outside specially-designed 'safe' areas.

Screw-cap lamp holders must be protected so that it is impossible to touch the screw cap so long as it is in contact with the female thread on the holder because there are a number of ways in which this may be alive. Thus on Figure 13.4 (b) the skirt must be sufficiently deep to prevent the cap being touched until contact is broken. In large fittings a diaphragm is often used instead of a skirt, Figure 13.4 (c).

Some lighting fittings have always presented a problem of over-heating, which may damage the flexible cord just above the lamp holder. This was investigated by P.D. Morgan for the ERA as far back as 1933[4], and his paper is still of interest. Subsequently there was considerable improvement in the design of large industrial lighting fittings, by means of good ventilation and provision of getting the heat away from the lamp cap and contacts. In later years the trouble returned, particularly in rather smaller fittings (100 W and 150 W or lower) following the introduction of the smaller and more efficient mushroom-shaped lamp bulbs. Not only do these get hotter but their smaller size allows lamps to be inserted in fittings which were designed for lower wattages, e.g. 100 W bulbs in 60 W fittings.

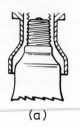 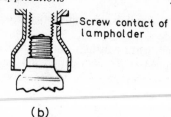

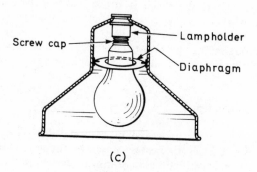

(a)

(b)

(c)

Figure 13.4 Protection of screw-cap lamps
(a) Lamp fully inserted
(b) Lamp cap still live but inaccessible
(c) Diaphragm guarding screw cap

In 1962, Niedle[5] stated that the maximum operating temperatures of metal springs should be

Phosphor bronze	135°C
Ordinary steel	170°C
Special steel	200°C

From this it seems clear that bronze contact springs at least could give trouble, and weakened springs will cause bad contacts and increased heating.

Flexibles and other cables (especially in *old* installations) should be carefully examined. Rubber hardens and crumbles while p.v.c. 'flows' when overheated so that short circuits become more likely. This may cause shocks if metal fittings are not properly earthed or if the cable fails at the cord grip of a pendant. The trouble is perhaps most likely to arise in enclosed fittings for damp atmospheres (e.g. bathrooms).

There have been improvements in design and materials but cables must obviously be regularly inspected and changed as necessary, particular attention being given at the cord grip and terminals. This seems likely to be a recurring problem for some years to come.

13.11 Bench lamps and spotlights on machines (bracket lights)

When lighting fittings can be handled in use they should preferably be supplied at very low voltage such as 12 V or 25 V from a safety isolating transformer made to the appropriate British Standard. Otherwise they must be effectively earthed. Preferably the lamp holder of a bracket lamp should be of insulating material and screened from mechanical damage by the shade. This is also necessary because the lampholder may not be strong enough to support a shade and may fracture near the base if the shade is struck. (Robust lampholders should be used which may be ceramic). Screw-cap lamps are not generally suitable for this purpose.

The same considerations apply to pendant lights which should not have switched lampholders except at very low voltages[9].

13.12 Portable standard lights

The same recommendations apply as for bracket lights. Where high powered lights, e.g. 500 W are necessary, screw-cap lampholders are required. With such loads, difficulties may arise at very low voltages and a supply of 110 V may be necessary. This should be from a safety isolating transformer with the centre point earthed and care must be taken with the cable connections and earthing.

Portable lamps often have long trailing cables and may be used on construction sites where they are subject to rough usage and abuse. The best quality heavy duty tough rubber or p.v.c. sheathed cable should be used and it should be protected against damage so far as possible.

13.13 Handlamps

Unless portable handlamps are of the safety type and contain no metal which can become alive in the event of a fault they must be earthed. Safety-type handlamps should always be used.

Because of possible misuse or abuse and use in confined or damp situations handlamps are usually supplied at a very low voltage (12 V or 25 V). It is sometimes found however that a supply of 50 V is used with the centre point earthed (i.e. 25 V to earth).

Handlamps should be supplied from a safety-type isolating transformer. No metal parts such as the guard should be or become in contact with a metal lampholder. In practice the requirements of Regulation 13 are difficult to meet and should, wherever possible, be avoided by using all-insulated equipment and/or very low voltages. The

safety type handlamp was 'invented' by Scott Ram, the first of HM Electrical Inspectors of Factories in 1900.

13.14 Plugs, sockets, extension leads and switches for portables and transportables

Domestic-type plugs and sockets are only suitable for light (e.g. domestic) industries and even in the clothing industries industrial types are preferred. In the UK these should be of a type covered by British Standards.

The types of plugs and sockets available are many and varied. For example, these include two-pin with no earth wire, two-pin and earth, 3-phase and earth or three-phase with circulating current earth monitoring. There are other types which are required for special purposes such as for heavy currents for vehicle battery charging, or waterproof plugs and sockets for dockside travelling cranes. All these types have a legitimate purpose and special requirements. Plugs for telecommunications and control circuits may have many small pins.

To ensure that the wrong pin shall not fit the wrong socket with all these varieties is important. For example, for the earth pin of a two-pin and earth plug to go into a live socket on a three-phase supply is very dangerous and has caused fatalities. This depends on differences in diameter and spacing and with all the varieties and possibilities single millimetres count.

On one occasion it was found that a rubber plug to one specification was sufficiently pliable to be *forced* into a socket of another type, with the earth pin in a live tube. This occurred when a contractor brought his own tool into a works and was not familiar with the plugs and sockets in use there.

The lesson from all this is that old installations remain with thousands of sockets for uncommon purposes. Great care must therefore be taken in purchasing plugs and sockets and an eye kept on 'foreigners' brought in by contractors. *The selection of such equipment must not be left to a non-technical purchasing agent or clerk.*

The construction is also important. The live pins and sockets must make contact *after* the earth pin. There must be an effective cord grip which does not damage the cable. For pliable armoured cable there must be an effective armour clamp and for heavy cables, a gland to grip the protective sheath firmly.

Internally it must not be possible for the earth wire to come loose and touch any of the live terminals (including the neutral). This is usually effected by the provision of fixed insulating separating barriers. When wiring a plug it is preferable to leave a little slack on the earth

connections so that if the cord grip fails, the earth wire breaks last. Terminals should be designed to grip without cutting into the conductor. Loosening of the screws holding parts of the plug together must not also loosen the grip on the conductors, particularly the earth wire.

The internal terminals must be clearly marked so that the conductors can be connected correctly. They are identified by their colours:

for single phase   Earth   — Green and yellow or green
                     Live   — Brown
                     Neutral — Blue

These colours are now widely accepted internationally but care must still be exercised with imported equipment. At one time red was used for the *live* conductor in the UK and for the *earth* in Germany and elsewhere and such cables probably still exist.

13.15 Portable tools (see also Chapter 4)

Hand tools should wherever possible be supplied at not more than 50 V or 110 V with the centre point earthed. They should be supplied from a safety isolating transformer. Unless they are completely encased in insulating material or double-insulated hand tools must be earthed. Special care is necessary in the maintenance of double insulation which is discussed in detail in Chapter 4.

s 2.18   Portable tools should be tested and inspected regularly and, for this purpose, a register of tools and the results of such tests and inspection should be kept and of any repairs carried out. These tests should include a test of the continuity of the earth wire preferably made with a substantial current at a very low voltage. There is a large number of *s 8.3.2* suitable instruments available for this purpose.

Figure 13.5 shows typical potential hazards relating to earth fault paths when using an electric drill.

13.16 Transportable machines

The same general recommendations apply as for portables. It is sometimes said that moderately large motors or conveyors, stackers etc. are unsuitable for a 110 V supply, but since 100 ton cranes have operated at this voltage in the past the objections cannot be sustained.

Transportable machines require thick cables which are liable to damage and, for these appliances, colliery-type trailing cables are probably the most suitable. Pliable armoured cable is often the best

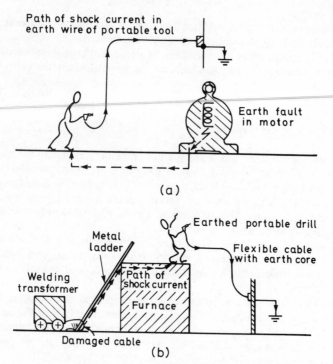

Path of shock current in earth wire of portable tool

Earth fault in motor

(a)

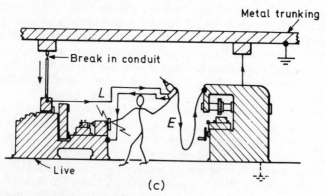

Welding transformer

Metal ladder

Earthed portable drill

Flexible cable with earth core

Path of shock current

Furnace

Damaged cable

(b)

Metal trunking

Break in conduit

L

E

Live

(c)

Figure 13.5 Unexpected hazards
(a) Shock caused by potential gradient in ground
(b) Shock from fault in 250 V supply cable in welding set
(c) Accident caused by improper connections to electric drill

(Reproduced from 'Electrical Accidents' by permission of the Controller of Her Majesty's Stationery Office)

but it requires specially constructed armour clamps and glands at the ends. Heavy duty plugs and sockets are also required.

s 8.3.10 For transportables, a circulating current earth proving circuit is strongly recommended (see Chapter 8) of which several types are available.

13.17 Cranes, excavators etc

The principal danger which arises from cranes is from overhead trolley wires. These must be guarded against contact by men entering the crane cab, working above floor level, e.g. millwrights and painters, or carrying long objects such as pipes or metal ladders, or on top of stacked material in a store. The diagrams in Figure 13.6 show ways in which this may be achieved.

s 9.3.1

s 1.1 For some work, however, particularly by electrical fitters, the conductors must be approached and sometimes the guards removed. At such times power must be cut off and isolators locked open. Generally it is advisable to operate a permit-to-work system in all but small works. Particular care must be exercised in the supervision of outside contractors, such as painters and glaziers who must comply with the regular works safety routines. There should be someone personally responsible for the safety of outside contractors to see that they do not wander into danger.

s 8.3.10 Large tracked cranes and machines such as excavators have trailing cables and are sometimes supplied at high voltages, e.g. 3000 V or 6000 V with virtually a small substation on board. This cable needs protection and colliery-type pliable armoured cable is recommended, which must have properly adjusted armour clamps and glands. It should be protected against short circuits and against earth faults by over current and earth leakage protection at the supply end. Earthing the machine is vital and circulating current monitored earth leakage protection is important. Heavy duty sockets and plugs with armour clamps and glands are necessary, or permanent connections. In all these matters colliery practice may be followed.

Cranes running on rails at ground level may be supplied by trolley wires but cable is often more convenient. A suitable arrangement is a revolving spring-loaded or mechanically-driven cable drum located on the crane and for the cable to rest in a trough for protection.

s 13.8

ss 9.4;

9.5; 9.8 Contactor panels on cranes, excavators, etc must be treated as open switchboards and the appropriate clearances provided. Cranes and similar machines are commonly provided with a crane protective panel in which contactors and various protective devices are located. Interlocks and over-run limit switches are necessarily connected to the live

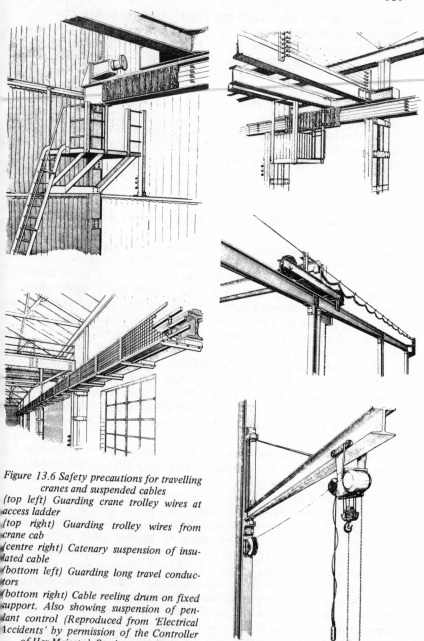

Figure 13.6 Safety precautions for travelling
cranes and suspended cables
(top left) Guarding crane trolley wires at
access ladder
(top right) Guarding trolley wires from
crane cab
(centre right) Catenary suspension of insulated cable
(bottom left) Guarding long travel conductors
(bottom right) Cable reeling drum on fixed
support. Also showing suspension of pendant control (Reproduced from 'Electrical
Accidents' by permission of the Controller
of Her Majesty's Stationery Office)

side of the main contactor and as a result some cross trolley wires on the crane remain alive when the main contactor is open. For this reason an isolating switch must be provided in the crane cab to make the whole installation dead for safe work and access to motors, etc. Where the crane cab is mounted on the crab there must be an isolating switch at the point of pick up. Apart from this there must be the usual isolating switch at floor level. This duplication is essential to safe working and maintenance.

A considerable amount of valuable guidance on safe electrical installation is contained in British Standards, for example BS 466: Overhead electric travelling cranes.

13.18 Work on construction sites and other temporary installations

Attention must be drawn to the comparatively large number of accidents on construction sites. This arises largely from the rough conditions and the temporary nature of many of the installations. Accident figures for one year are analysed in Table 13.1. In past years portable appliances caused the greatest number of accidents (including fatalities) but recently there has been an improvement. This is probably due to voltage reduction and the increasing use by workmen of shoes or boots with moulded plastic or rubber soles.

Most large contractors supply portables at 110 V c.p.e., or better still at 50 V c.p.e. This is a condition of contract for work on many power-station sites and for some gas boards. All-insulated or double-insulated tools cannot easily be made sufficiently robust for this duty, although, rather surprisingly, a double-insulated road drill has been designed.

ss 13.12 to 13.16

Fixed or heavy transportable equipment can usually be efficiently operated on 3-phase 110 V supplies, as stated above. The cables supplying them must be very robust and further protected where necessary, bearing in mind the possible irresponsible or unplanned use of earth-moving equipment. Double-wire-armoured cable is therefore recommended for buried cables (it is more flexible than steel-tape-armoured) and mining-type pliable armoured cables for trailing cables, as recommended in section 13.16 on transportable apparatus. On the largest jobs, however, such as for tunnelling and for the heaviest earth-moving equipment (e.g. for a canal or major reservoir), machines may necessarily be powered by, say, 3 kV or 6 kV motors. Here, earth-leakage protection and monitored earthing should always be used. (It must be remembered that work of this type is usually subject to the Factories Acts and the Electricity Regulations). Electrical equipment for fixed and

s 13.12

s 13.3

s 13.16

Table 13.1
ACCIDENT FIGURES FOR CONSTRUCTIONS SITES

	Building		Construction Engineering	
	Fatal	Total	Fatal	Total
Portable tools (class 1)	–	1	–	1
Lamps	–	1	–	–
Testing sets, including lamps, instruments and test leads	–	–	1	1
Plugs, sockets couplings and adaptors	–	–	–	1
Cables and flex for portables (other than test sets)	–	2	1	3
Electric hand welding (excluding welding eye flash)	–	–	–	1
All other portable apparatus (including pendant control)	–	3	–	2
Transformers and reactors	–	–	–	1
Other switch, fuse and control gear above 650 V	–	–	–	1
Circuit breakers, not exceeding 650 V	–	–	–	5
Contactor and other control apparatus below 650 V	–	1	–	3
Switches and links not exceeding 650 V	–	–	–	5
Fuse gear not exceeding 650 V	–	–	1	9
Fixed lamps	1	2	–	1
Cables and accessories (excluding cables, flexibles etc for portable apparatus and buried cables)	–	11	–	16
Buried cables	–	39	–	36
Contact by cranes and similar machines	4	5	2	5
Direct contact by persons, materials, tools	3	4	–	–
Batteries	–	–	–	1
Radio, TV, electronic instruments and power packs	–	–	–	1
Apparatus not classified	–	–	1	1
Total	8	69	6	94

mobile cranes with trailing cables is specified in the relevant British Standards.

Except on small jobs that are quickly completed, where it may be taken from a fixed installation, lighting should be carried out at not more than 110 V c.p.e., with portable lights at not more than 25 V to earth. This again is written into many contracts.

The main danger from cables is from stray temporary-lighting circuits. These must be adequately protected from damage. Tough-rubber-sheathed or equivalent cables should be used, but this alone may not give sufficient protection, and, for places where the supply is needed only for a very short time, substantial additional protection may be obtained by threading cables through tough garden hose.

The installation and protection of switches, fuses, plugs, sockets and their transformers on a building site has always presented considerable

difficulties, and the Building Research Station at Garston has developed prototype protected portable units for this purpose. These are now covered by BS 4363, incorporating sockets meeting BS 4343.

The London Master Builders' Association have issued a very useful Bulletin (No 15) for the guidance of their members, and the whole matter is covered by CP 1017: 'Distribution of electricity on construction and building sites', which should be consulted.

13.19 Overhead lines and supply cables

All the above risks are common, and they occur in abnormally hazardous situations. There is, however, one special class of risk which is peculiar to constructional work, namely accidental contact with overhead conductors. The types of accident which may happen are clearly illustrated by Figure 13.7 and 13.8. These have caused fatalities, and they are typical of a great many more. Table 13.2 is a typical 5-year summary of accidents.

Table 13.2
SUMMARY OF OVERHEAD LINE ACCIDENTS

Year	Contact by cranes or similar machines		Direct contact by person or material and tools		Totals	
	Fatal	*Total*	*Fatal*	*Total*	*Fatal*	*Total*
1957	–	–	1	7	1	7
1958	8	14	1	4	9	18
1959	6	8	4	8	10	16
1960	2	7	6	14	8	21
1961	4	16	1	1	5	17
Total	20	45	13	34	33	79

The degree of irresponsibility shown by some contractors and their staff is illustrated by a fatal accident which occurred during the laying of a pipeline in open country. It was to pass near a major transmission line, and the Area Board warned the contractors of the danger and said that, if they were given a few days warning and the line could not be made dead, they would, without charge, supply a linesman to watch the work of the crane, to see that it kept out of danger. The contractors did not bother.

To divert an overhead line requires months of planning. A route must be surveyed, wayleaves negotiated, the consent of the Minister

(a) Crane rope severed by arcing

(b) Contact with line by tipping lorry

(c) Contact with line by mobile crane jib

(d) Use of metal or metal-bound ladder

(e) Bricklaying

(f) Handling steel scaffold pole

Figure 13.7 Typical overhead-line accidents
(Reproduced from 'Electrical Accidents' by permission of the Controller of Her
Majesty's Stationery Office)

344

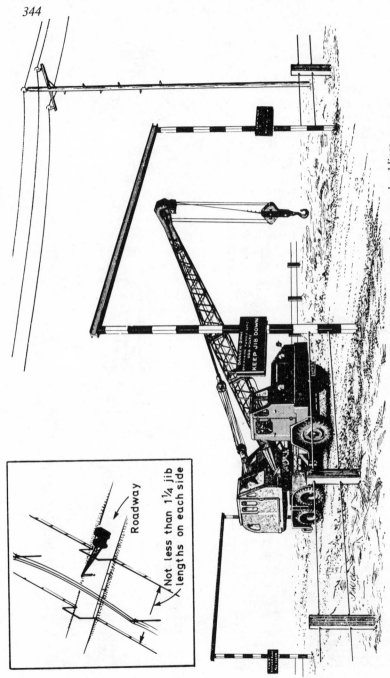

Figure 13.8 Post and wire fences and 'goalposts' guarding approaches to overhead lines

(Reproduced from 'Electrical Accidents' by permission of the Controller of Her Majesty's Stationery Office)

obtained after he has consulted the Post Office, the Ministry of Defence and the local planning authority. Also, materials must be requisitioned or purchased and the job fitted into a tight construction programme. All this takes time.

It is quite useless to ask for a transmission line to be put underground or diverted at short notice. To divert a major transmission line is *extremely expensive*, and considerable notice may be needed even for it to be made temporarily dead.

On one occasion, after a road plan had been prepared and the Area Board informed, it was found that an 11 kV cable passed right under a house. The plan had been changed without the supply authority being told. Possibly, the gas mains were also under the houses. It is by no means unknown for builders to be found laying bricks or erecting scaffolding literally within inches of a live line.

Because of the seriousness of some of these matters, the Construction (General Provisions) Regulations contain the following special requirements for such work. These are in addition to the Electricity Regulations, which also apply; the UK Factory Inspectorate takes a very serious view of accidents caused by any breach of these regulations.

Regulation 44: (i) Before any operations or works to which these Regulations apply are commenced, and also during the progress thereof, all practicable steps shall be taken to prevent danger to persons employed from any live electric cable or apparatus which is liable to be a source of such danger, either by rendering such cable or apparatus electrically dead or otherwise.

(ii) Where any electrically charged overhead cable or apparatus is liable to be a source of danger to persons employed during the course of any operations or works to which these Regulations apply, whether from the operation of a lifting appliance or otherwise, all practicable precautions shall be taken to prevent such danger either by the provision of adequate and suitably placed barriers or otherwise.

13.20 Rubber gloves, shoes and mats and insulated tools

There are British Standards for rubber gloves and rubber mats for electrical purposes. There are, however, severe practical limitations to their use. To be reliable, rubber gloves must be very thick and tough. In the case of insulated tools it is necessary to ensure that the thickness of the insulating layer is uniform and has no thin patches. These matters are, rather surprisingly, best dealt with in the British Standard for firemen's axes with insulating handles, which should be consulted as to the manufacture and routine testing.

Rubber boots and shoes may also present a problem. Welders commonly wear rubber wellingtons, and these undoubtedly contribute to safety, although how many accidents they have prevented cannot be estimated. Recent statistics, however, indicate that the great increase

in the use of moulded tough plastic (including rubber) working shoes has significantly reduced the number of fatal and other accidents from the use of portable tools. One man wearing such shoes had been working for some time when he handed his tool to his mate, who was immediately electrocuted. The mate was wearing ordinary leather shoes. Such boots and shoes, however, cannot by *themselves* be accepted as a satisfactory safeguard against shock, since the protection offered may fail with wear.

A warning must be given with respect to rubber mats and shoes. These may be made of *conducting* rubber specially prepared to dissipate electrostatic charges in hospitals, chemical works or explosive works. If they are made according to the relevant British Standards, they should be marked, but this mark may not be recognised or understood. This is an additional argument for having a British Standard for insulating shoes and boots.

BS 3825 covers the construction of electrically *conducting* rubber footwear but a British Standard for *insulating* footwear would be valuable.

13.21 Conducting rubber flooring and similar materials

Conducting rubber is sometimes used for flooring, 'rubber sheets', belting, shoes, etc to effect rapid dissipation of electrostatic charges in such places as explosive works and hospital operating theatres.

A warning must be given about this matter. If the resistance is too low, in the event of conducting rubber coming into contact with live conductors associated with power supply, lighting, instruments etc, sufficient leakage current may flow to heat up the material and it may eventually catch fire. They do *not* provide protection against electric shock.

13.22 Electric arc welding

There are three main hazards associated with electric welding, namely electric shock, conjunctivitis or arc-eye and fire.

Although hand welding is commonly carried out at a relatively low voltage e.g. approx. 80 V a.c. on load with approx. 110 V a.c. on open circuit, the live welding electrode is necessarily exposed and occasional fatalities occur in difficult situations such as in tanks or the double bottoms of ships. The welding pliers or electrode holders should be encased in insulation so far as is compatible with their use and the dissipation of heat. But this is insufficient and the welder depends in

part on the good condition of his gauntlets, which must be unabsorbent heat resisting and as pliable as possible.

Conjunctivitis, which is described in Chapter 3 is prevented by the proper use of a helmet or face mask which is fitted with dark glass to facilitate work. Assistants to welders and others in the vicinity may also need protection. Goggles with thick clear glass give some protection and permit the work to be seen but special glasses giving both a clear view and good protection are available. *s 3.3.2*

Apart from direct ignition of gas or flammable material by the welding torch, fires may be caused by poor earthing of the workpiece and/or inadequate continuity of the return circuit to the welding transformers or generator. This presents a number of technical problems, but if it does not receive attention, stray current returning by unauthorised paths may cause fires, burn out the earth wires of portable tools, and upset telecommunications, control equipment, etc and damage instruments. One difficulty is that operators may not have the technical knowledge to appreciate the significance of lack of care and this good supervision is necessary. One incident was traced to some makeshift arrangements for electric welding 200 yards away. The welding return circuit should be earthed at the job rather than at the welding transformer, but there must be an adequate and preferably insulated conductor between these points. *s 8.3.2*

All these and other matters are discussed in detail in Safety Health and Welfare Pamphlet No. 38 'Electric Welding'[6], which should be consulted. Note that in the UK, electric welding, being an 'electro thermal process' (subject to certain provisions which are very important) is exempt from some detailed requirements of the Electricity Regulations. The Memorandum on The Electricity Regulations should be consulted.

13.23 Electrical testing

Safety in electrical testing depends primarily on the skill of the operators unless it is carried out at a very low voltage. Because the work is very variable no detailed regulations could be drawn up which would be widely applicable. For this reason testing and research in the UK are also (subject to certain important conditions) exempt from some detailed requirements of the Electricity Regulations.

Apprentices must learn testing, but while they are learning they must be very carefully supervised to ensure they only undertake work for which they have received adequate practical and technical training for them to work safely. These matters are subject to a special amendment to the Electricity Regulations. *ss 1.1; 3.2*

Some routine testing during manufacture is carried out by semi-skilled or technically unskilled persons. Unless this done at very low and completely safe voltages, it must be carried out in 'tunnels' which may be of transparent but robust plastic, or in locked and interlocked cabinets, so that it is impossible to obtain access to conductors while they are alive.

For other work, particularly at high voltages, skilled technicians and engineers must have access to enclosures for connecting up and for making adjustments, but steps must be taken to see that the enclosure is clear before the power is switched on. Typical examples are testing transformers, circuit breakers and cables. These usually require locked enclosures with all access doors interlocked with the supply input switch. Because such enclosures may be very large, it is often usual to have a gallery, bridge or other vantage point from which the whole space can be seen. A point from which dangerous tests may be viewed in safety is particularly important for circuit-breaker testing. All such

Table 13.3
ACCIDENTS IN ELECTRICAL TESTING IN 1974

Apparatus being tested or used in testing	No. of accidents
TESTING ON ELECTRICAL SYSTEMS	
Portable tools: Class 1	1
Testing sets, including lamps, instruments and test leads	7
Plugs, sockets, couplings and adaptors	4
Contactor and other control apparatus below 650 V	7
Switches and links not exceeding 650 V	5
Fuse gear not exceeding 650 V	1
Cables and accessories (excluding cables, flexibles, etc. for portable apparatus and buried cables)	1
Fixed test apparatus and their cables	2
Batteries	1
Radio, TV, electronic instruments and power packs	4
Apparatus not classified	1
Total	34
ELECTRICAL TESTING OF PRODUCTS	
Heating and irons	1
Testing sets, including lamps, instruments and test leads	3
Plugs, sockets, couplings and adaptors	1
Cables and flex for portable (other than test sets)	1
Rotating and electrical machines	1
Circuit breakers not exceeding 650 V	1
Fixed test apparatus and their cables	4
Radio, TV, electronic instruments and power packs	2
Total	14

work must be carried out in accordance with a carefully prepared and detailed routine or code.

A third aspect is the necessity of providing a safe situation for testing, repairing and adjusting small instruments and apparatus while it is alive, for example electronic equipment such as television sets or radar. Only very skilled and experienced persons may do such work, it is normally carried out in an 'earth free' area. The latter introduces a number of detailed technical requirements some of which are set out in Safety Health & Welfare Pamphlet 31 'Safety in Electrical Testing'. The subject of interlocking is dealt with in detail in Chapter 4, and in Safety Health & Welfare Pamphlet 24.

13.24 High frequency furnaces and h.f. heating

H.F. processes introduce several unusual hazards. Three types of equipment are used, the first type is the high frequency steel making furnace which may be of perhaps only a hundredweight capacity, or a few *ss 3.2.7* pounds in a laboratory, but can be as large as several tons capacity. *to 3.2.12* Except for small laboratory use these furnaces are usually supplied from a high frequency induction generator. Hazards arise in the control enclosures where, in addition to the usual precautions, access to high voltage contactors must be prevented while the main power is on but allowed when the main power is off but low voltage (e.g. 50 V a.c.) is on for the control circuits. It must, however, be possible to view con- *ss 9.7; 9.8,* tactors in operation, through a window, without having access to them. *9.9*

The other special hazard arises from the high voltage, water-cooled, inductor coil in the furnace itself. When a furnace lining fails, if the molten steel reaches the inductor it may in certain circumstances be punctured, allowing a jet of water under high pressure to penetrate into the melt, causing a violent and dangerous explosion. This phenomenon was at first very obscure but was eventually solved and was prevented by a high speed earth leakage relay and an electrode embedded in the refractory furnace lining; but there remain some technical problems.

At the other extreme are the radio frequency furnaces and heaters used for heating plastics and setting synthetic adhesives and similar purposes. These were mostly supplied from large thermionic valve *ss 9.7; 9.8;* oscillators after the manner of high power radio and TV trans- *9.9* mitters, but with the development of solid state electronics they may be powered by static equipment. The risks here arise primarily from access to high-voltage equipment for servicing and adjustment, and the conditions resemble those for steel melting or electrical testing.

The shock risk from contact with the applicators is not so great *s 3.2* because of the very high frequency (see Chapter 3) but it must not be

overlooked. There are, however, considerable difficulties in earthing casings and enclosures of such very high frequency equipment because of standing waves and the high impedance of conductors.

May's *Industrial high frequency electrical power*[7] is a useful source of background information but the whole problem of high frequency heating equipment was covered in *Safe usage of induction and dielectric industrial heating equipment*[8], which is in effect a code of practice drawn up in 1962 by BEAMA in consultation with HM Factory Inspectorate following a number of accidents and failures.

13.25 Office equipment

Some people may be surprised to learn that this may be dangerous but fatalities have occurred. Today an office with computers, electric typewriters, photocopiers, addressing machines and so on may contain a large amount of electrical equipment. This is dealt with in some detail in BS 3861, Specification for Electrical Safety of Office Machines, Part 1 deals with 'General requirements and tests of earthed equipment'.

13.26 Planned inspection and maintenance (see also Chapter 2)

Some statistical aspects of planned maintenance have been dealt with briefly in Chapter 2. These are relevant to the managerial aspects of planning in a large organisation.

It is best to forestall failures if possible in the interests of both safety and economy and this requires careful planning. For example, the earthing of all equipment should be regularly checked, and this must be done much more frequently for portable apparatus than for fixed plant.

All planned inspections and maintenance should be based on a register to ensure that nothing is omitted; in this the results of insulation and earth loop impedance tests should be recorded and all repairs, whether arising from regular inspection or random faults. The register should be regularly examined by a person with the authority to initiate any necessary action and certainly not below the rank of a senior foreman except in small establishments. It will quickly show up unsuitable equipment which is giving excessive trouble and should probably be replaced by better designs. This ensures an efficient installation as well as a safe one. Accidents and breakdowns can be very expensive.

There may be considerable differences of opinion about routine testing of some apparatus. Most heavy electrical equipment may be tested without difficulty, e.g. the condition of the oil in transformers and switchgear, to see that it is free of inflammable gas or vapour in solution and of water and carbon etc, and it can often be cleaned and re-used. The settings of protective devices can also be checked, e.g. by current injection, but care is necessary to ensure that any temporary connections or other 'alterations' are restored to normal.   s 10.15

With low current equipment the position is different; at one time it was usual to check the mechanical features of telecommunication equipment for adjustment at regular intervals, but it was often found that the disturbance tended to increase the probability of malfunctions. Some details can, however, be tested for 'drift' or gradual deterioration of characteristics without disturbing them. Solid state devices may be very sensitive to over-voltages and sometimes to over-current (see Chapters 6 and 7) and care is necessary and some test apparatus must not be used. It is very important to follow the maker's instructions.

A medium-size works may well not have adequately trained specialists for some maintenance work, e.g. servicing high voltage circuit breakers, but in some places the electricity board is willing to do this for its customers on a routine contract basis. I have investigated several accidents caused by good maintenance electricians coming to grief through gaps in their experience, for example one man caused a flashover and explosion because he did not know that it is dangerous to open circuit the secondary of a current transformer while the primary is loaded. Also works electricians may not understand the possibly devastating results of a flashover or short circuit on the lower voltage side of a large transformer.

Similarly it is sometimes felt that regular inspection and testing of the insulation levels, earth loop impedance and condition of motors etc either takes up too much of the time of skilled staffs or that they will not do conscientiously what they consider a dull and non-urgent task. This also can be done by contract and some engineering insurance companies employ inspectors for this work. This has the advantage that they will not be inspecting their own work.

Some maintenance work must be done on 'live' gear if a factory is not to be brought to a standstill. This can often be done with a high degree of safety but *it requires very special training and equipment, specially designed techniques, and close supervision.* The electrical supply authority may be able to undertake such work.

It may not be generally known that the low tension cable jointing of services has for many years been carried out with the cables alive and I have not heard of an accident (though I think there must have been a few). In fact, on one occasion in the Midlands, owing to a mistake, a

jointer successfully connected a 430 V service to a live 3000 V cable — fortunately it blew up the consumer's service cut-out before anyone was hurt.

13.27 Telecommunications; radiation hazards

The hazards associated with operating and maintaining high-power T.V. and radio transmitters are not unlike those for high-frequency heating and testing. D.H. Shinn[10] has comprehensively reviewed this subject. The reader is also referred to the advice given in sections 13.23 and 13.24 which will be helpful in this connection.

13.28 Textile and clothing industries

The chief centres of danger are in the preparation and spinning of the material prior to weaving, which is generally accompanied by the evolution of a great deal of fluff, etc. In tests, it was found that cotton fluff was easily ignited by a short circuit at the cord grip of a pendant light as was fluff from artificial-silk rag and waste. Oiled pure-wool fluff did not ignite very easily except when mixed with a small amount of cotton. The presence of artificial-silk braiding from twisted flex was, however, sufficient to get it well alight. Clean wool and real silk did not fire.

The short-circuit currents were extremely small, and with an overfused circuit and heavier short-circuit currents, ignition would be much easier. Ignition may also take place on open or cleat wiring or at exposed cable and makeshift joints once frequently found between the end of a piece of conduit and the entry to a motor or starter and, in fact, very serious fires have been started in this way.

13.28.1 Preliminary processes and spinning

Since the textile processes are rather specialised, the following notes may help to make the application of these results clear. Taking cotton as the standard, the raw cotton is first broken up and opened by means of a number of special machines. This is essentially a dusty and fluffy process. After being thoroughly broken down it is passed to the carding machines, which tear apart the filaments of cotton and deliver it as a light, almost impalpable, web which may be described as a large amount of air held together by a very small amount of cotton. This is the most fluffy and ignitable stage of the whole proceedings. From here the

'sliver' or web of cotton filaments is gradually reduced and pulled out until finally it is passed to the spinning, doubling and weaving departments in which the dust risk is comparatively small.

The progress from the raw state of wool to woollen yarn is very much the same, although the machinery tends to be of a bigger and heavier nature. There is the essential difference, however, that whereas cotton filaments are extremely flammable, dry wool is difficult to ignite and will not normally continue to burn unless in the presence of some artificial draught. For certain kinds of cloth, however, oil is introduced into the wool at an early stage and it thereby becomes, to some extent, flammable. The risk is still by no means as large as with cotton, although fires have occurred in woollen mills.

There are also two sub-industries, one auxiliary to the cotton trade and the other to the woollen trade, in which the risks appear to be very much greater than in either of the pre-spinning processes mentioned above. First there is the process of rag grinding and the manufacture of shoddy, mungo or flock. Although these — with the exception perhaps of flock — are part of the woollen industry, the basic material may be wool, cotton, silk or artificial-silk rags or waste wool, cotton or artificial silk. After grinding or breaking up, the material, which has generally been impregnated with oil, passes through a series of processes corresponding to the pre-spinning processes mentioned above, but owing to the use of a proportion of cotton and artificial silk the danger is apt to be much greater than with pure wool. Secondly there is waste-cotton spinning. This material is generally of very short staple and the dust or fluff risk is correspondingly increased; in fact the opening and carding of waste cotton and the grinding and garnetting of artificial-silk rag appear to produce the highest risks of any of the pre-spinning processes.

A reference should also be made to 'fibro', which is artificial silk cut into short filaments to adapt it to the standard cotton and woollen spinning machinery. (Cheap knitting yarn is often made from waste fibro, not wool.) The standard of risk here is that for cotton or artificial silk, not as for wool, which the yarn may resemble superficially.

The obvious precaution is to have a totally metal-enclosed installation, and this is already adopted to a large extent in the Lancashire cotton industry, where the risk of fires at the cord grip is well appreciated.

13.28.2 Weaving

There are no particular risks in the process of weaving, except in the case of jacquard looms used in the lace and carpet industries and for a

few other purposes. The strings and cords may catch fire and give rise to a general conflagration. The obvious cure is to use non-flammable materials.

13.28.3 Carbonising

There is one process in connection with woollen manufacture which is anomalous, i.e. carbonising, of which there are two types. In both of these the woollen fibre, fabric or rag is treated with sulphuric acid to destroy any cotton or other vegetable fibre, which is thereby reduced to a friable dust. One of the two processes referred to above is employed with rags and what may be referred to as raw materials, and the second with piece goods such as blankets. Two or three fires arising in the carbonising of blankets have been investigated, and are stated to have started at the cord grip of pendant lamps. The process did not appear to be particularly dusty of fluffy, and it may be that it was the flexible cord itself which caught fire, possibly having been affected by acid fumes, although the dust mentioned above may also have been involved, but the effect of acid fumes on insulation and metallic sheathing must be considered.

13.28.4 Making-up

In making-up textiles into garments, particularly artificial silk, the chief danger arises from open portable radiators which are not infrequently used where workrooms are of a small, improvised type, and from electric irons which have been known to burn through a wooden table top. Sometimes a tell-tale lamp is used to indicate when the irons are left switched on. A good deal of protection can be obtained by use of the special iron rests. One may feel, however, that it would be an added precaution and a justifiable measure of economy to have the power switched off by a time switch at the hour at which the workrooms normally close, as fires of this description usually occur when the premises or department concerned are unattended.

Any flexible cable, which has a flammable braiding or packing is to be deprecated whether it is asbestos-wrapped or not. Plain t.r.s. or PVC sheathed cable or t.r.s. cable with hard-thread braiding moulded on to the rubber has proved satisfactory.

13.29 Undervoltage and phase-failure

In conclusion it is important to mention a point which has not been covered elsewhere. This is the effect of a serious fall in voltage without

an actual cessation of supply. This is met in conventional control equipment by using low-volt releases instead of no-volt releases, but it may have secondary effects which can be overlooked. Some apparatus may not stop or slow down but draw extra current from the mains causing overloading. An associated trouble may occur when the supply to one phase of an induction motor fails, e.g. because of a blown fuse or loose connection. It may continue to run and in doing so, overload the other two phases. Special phase failure relays are available to overcome this trouble. Once stopped, the motor will not restart in the absence of special control features.

Low voltage can cause erratic operation of complex control systems if a few relays fail or are sluggish or operate in the wrong order. I have investigated a few failures caused this way and this possibility should *ss. 9.10* always be investigated when vetting complex systems. There may also *to 9.17* be a danger from this cause with solid state controls. I have found that the operation of a pocket calculator may become completely erratic when the battery voltage is low.

References

1. Hill, W., 'Development of a motor terminal box with phase separation', *AEI Engineering* **1**, 11 (Nov 1961).
2. Schwarz, K. K., 'The design and performance of high and low voltage terminal boxes', *Proc IEE*, **109**, Part A (April 1962).
3. Hill, H. R., 'Differential expansion of conductor terminations', *Elec. Times* (22 Jan 1959).
4. Morgan, P. D., Taylor, H. C. and Lethersich, W., 'The heating of domestic pendant lighting fittings and their connecting leads' *Jour. IEE* 73 pp. 545–568 (1933).
5. Niedle, M., 'Cables for lighting', *Elec. Times* (5 April 1962).
6. Safety Health and Welfare Pamphlet No. 38, *Electric Welding*, HMSO.
7. May, E., *Industrial h.f. electrical power*, Chapman & Hall (1949).
8. BEAMA 'Safe usage of induction and dielectric industrial heating equipment'.
9. Electricity Regulation No 13 (This regulation is far reaching and complicated. One effect is that it virtually prohibits the use of accessible metal lampholders for portables. See Memorandum on Electricity Regulations.)
10. Shinn, D.H., 'Avoidance of radiation hazards from micro-wave antennas', *Marconi Review* No 201 (1976).

Appendix –The Flixborough Explosion

The report of the court of enquiry into the disastrous vapour explosion at Flixborough in June 1974 was published in 1975. The explosion was the result of a calamity hazard such as I have already mentioned. There is no suggestion that the source of ignition was electrical, but it could well have been, and for this reason a number of extracts from the report which have a lesson for electrical engineers are given below.

'The immediate neighbourhood of the (Flixborough) works consists largely of farms and has a very low population density; the perimeter of the works is surrounded by fields. The siting of the works was fortunate, had they been in a densely populated area there is no doubt that death and serious injury would have been on a much greater scale and that the property damage, even if more localised, would have been more serious. The facts that 72 out of 79 houses in Flixborough, 73 out of 77 houses in Amcotts, 644 out of 756 houses in Burton-upon-Stather and 786 houses in Scunthorpe were damaged to a greater or lesser degree speak for themselves. [see Section 11]

In his evidence to us, Mr V.C. Marshall who was appointed Safety Adviser to the Transport and General Workers Union on the day after the disaster and who was called on their behalf stated 'hazard analysis recognises that hazard cannot be entirely eliminated and that it is necessary to concentrate resources on those risks which exceed a specific value'.

We agree with this statement which accords with reality. No plant can be made absolutely safe any more than a car, aeroplane, or home can be made absolutely safe. It is important that this is recognised for if it is not, plant, which complies with whatever may be the requirements of the day tends to be regarded as absolutely safe and the measure of alertness to risk is thereby reduced.

When Mr Marshall refers to risks exceeding a specific value we understand him to refer to risks which exceed what at a given time is regarded as socially tolerable, for what is or is not acceptable depends in the end upon current social tolerance and what is regarded as tolerable at one time may well be regarded as intolerable at another. Nowhere perhaps is this more apparent than in the field of road transport where the construction and use regulations have, over the years, become ever more stringent.

Finally, in these introductory paragraphs we would, albeit may appear only to be stating the obvious, point out that even when resources have been concentrated on an unacceptable risk that risk will not be completely eliminated. It can never be reduced beyond a certain point because not only may there be human error in the operation of equipment, there may also be human error in the construction of a safety device built in to the equipment. It is to reduce risk from such error that, for example, there are dual braking systems on cars. The existence of the second system substantially reduces the risk of dire consequences if the first system fails but no one can guarantee that the second system will not also fail.

[In Chapter 2 under 'reliability' I have suggested a method of identifying points where expenditure on increased reliability or duplication will be most effective.]

The disaster was caused by the introduction into a well designed and constructed plant of a modification which destroyed its integrity. The immediate lesson to be learned is that measures must be taken to ensure that the technical integrity of plant is not violated. We recommend:

(i) that any modifications should be designed, constructed, tested and maintained to the same standards as the original plant.

(ii) that all pressure systems containing hazardous materials should be subject to inspection and test by a person recognised by the appropriate authority as competent after any significant modification has been carried out and before the system is again brought into use.

[This could apply also to electrical and associated systems]

At the time of the installation of the by-pass the key post of Works Engineer was vacant and none of the senior personnel of the company, who were chemical engineers, were capable of recognising the existence of what is in essence a simple engineering problem let alone solving it. We consider that there are in this connection two important lessons to be learned:

(i) that when an important post is vacant special care should be exercised when decisions have to be taken which would normally be taken by or on the advice of the holder of the vacant post. This, in the present instance, would have involved the reference by senior management to Mr Hughes of the problems created by the defect in Reactor No 5 and the design of the by-pass.

(ii) that the training of engineers should be more broadly based. Although it may well be that the occasion to use such knowledge will not arise in acute form until an engineer has to take executive responsibility it is impossible at the training stage to know who will achieve such a position. *All engineers should therefore learn at least the elements of other branches of engineering than their own in both academic and practical training.*

Although unconfined vapour/air explosions have been known to happen in other parts of the world, there is a marked scarcity of information about the conditions under which an unconfined vapour cloud can result in an explosion or what is the mechanism leading to such an explosion. We do not know to what extent it is practicable to obtain this information but if it can be obtained it would clearly be useful.

[I have indicated above the limitations on existing recommendations for small leaks and the complete absence of knowledge on how to deal with major failures.]

Index

(The numbers in bold type indicate the more important references)